Agricultural Valuations
A Practical Guide

R G Williams FRICS, FAAV

2008

A division of Reed Business Information

Estates Gazette
1 Procter Street, London WC1V 6EU

First published 1985
Second edition 1991
Third edition 1998
Fourth edition 2008

ISBN 978-0-7282-0551-2

Cover design by Rebecca Caro
Image courtesy of Rex Features
Typeset in Palatino 10/12 by Amy Boyle
Produced by Elsevier Architectural Press

Contents

Foreword

Agriculture, like any other industry, continues to evolve and to change with ever increasing demands from politicians, the consumer and the environment together with the economic pressures that seem to be never-ending. This is certainly true since Gwyn Williams produced the last edition of the *Agricultural Valuations* 10 years ago. In particular the industry has had to adapt to the changes imposed by the introduction of the single farm payment system and every one has had to re-examine their businesses in the light of the changes from production support to the new system. This change was further complicated by the introduction of an area based support system in England with a completely different system in Wales and Scotland.

As a result of these changes, valuers businesses have also changed with a considerable growth in advisory work, particularly in relation to environmental schemes and rather less on traditional income streams. It is also 10 years since the introduction of the Agricultural Tenancies Act. This has altered the terms of occupation of large areas of the countryside.

Taxation issues have also been more prominent, considerable thought has now to be given before much advice can be given to our clients on a range of topics. In particular where we wish to retain agricultural relief with recent case law favouring the Revenue.

On behalf of the Central Association of Agricultural Valuers, I am delighted to recommend this book to all those involved in the profession, from students, probationers and qualified members. I believe that this new publication will again be essential reading to all who are involved in the professions in the countryside. I would also congratulate Gwyn Williams for his continuing excellent service to the profession.

P G Martyn FRICS FAAV
President CAAV 2007/2008, Beeston, Cheshire

Preface to the fourth edition

Ten years have elapsed since the publication of the third edition. Much has changed in the agricultural sector. With the passing of the Arbitration Act 1996, the Planning and Compulsory Purchase Act 2004, the Regulatory Reform (Agricultural Tenancies) (England and Wales) Act October 2006, the law relating to agricultural properties and tenancies has dramatically changed.

Also, the decisions in the *Antrobus* and the *McKenna* cases affecting inheritance tax has had a marked effect on tax liability in respect of farms and farmland. It therefore became necessary to update this publication.

I am indebted to the following for their help and assistance in preparing and amending various chapters — David Carter, Paul Wright, Tony Carver, Glenys Tucker, Geoff Coster, Michael Hardman, Tim Swallow, Alan Lane, Lee Ward, Stephen McLaughlin and Greg Christopher. Also two very competent and patient secretaries — Miss C Morgan and Miss C Pearson for typing and re-typing.

In particular I would like to thank the President of the Central Association of Agricultural Valuers, Graham Martyn, for writing the foreword and the President and Council of the CAAV for permission to make use of some of their specialised information on costings etc and, in particular, Jeremy Moody CAAV Adviser for checking the legalities quoted and for helpful advice.

R G Williams
Ross-on-Wye
May 2008

Preface to the first edition

This book started its life as a few articles dealing with specific types of valuation, written for the benefit of a number of young agricultural valuers, to assist in their professional examinations. I have been encouraged by several practitioners to develop these articles into book form and this publication is the result. Many young agricultural valuers have, over the last few years, had considerable difficulty in obtaining proper experience in agricultural valuations. The problems have increased as firms have become specialised, the work-load of principals has changed and increased, resulting in little time being available to instruct young valuers. I was further encouraged to publish this book as there is no similar, up-to-date publication, currently available.

Some valuers may consider that some of the figures given in the example claims are high and in some cases items may even appear to be duplicated. Nevertheless, it is the duty of any claimant's valuer to prepare a proper and full claim as he may otherwise be negligent, if he does not do so. Valuation is not an exact science and virtually every claim made by a practitioner is subsequently negotiated.

This book is essentially a practical guide drawn from more years experience in the profession than I care to remember. The views expressed and methods of valuation are, as I use, and interpret matters. Whilst every effort has been made to ensure the accuracy of the information given and opinions expressed, I cannot, however, accept any responsibility for any inaccuracy of these interpretations and valuation methods. I hope this book will prove to be of value as a practice guide to young and older valuers alike.

I would like to take this opportunity of expressing my appreciation to the President and Council of the Central Association of Agricultural

Valuers for allowing me to use some of their published costings and also their encouragement in writing this book.

My special thanks go to Mr Alan N Lane, FRICS, CAAV of Knutsford who has corrected the manuscript, written the article on pigs and also given me many valuable suggestions. Also to the following who have given valuable advice:

Mr Robert Thomas, MA, FRICS, CAAV Cowbridge; Mr Alan Brown, FRICS, CAAV Newport; Mr Malcolm Williams, CDA Ross-on-Wye and a leading compensation valuer.

I further wish to thank the following members of my own firm for checking the calculations, advice and making suggestions:

Mr Nigel Morris, ARICS, NDA, CAAV; Mr Michael Taylor, BSc, ARICS and my son Mr Richard Williams, BSc, ARICS, ASVA, CAAV.

Finally thanks go to four very patient secretaries who have typed and re-typed this work.

October 1984   *R G Williams*
Coles, Knapp & Kennedy
Ross-on-Wye

Table of Statutes

Table of Statutory Instruments

Table of Cases

History

1.1 Agricultural valuations, as commonly known, can be said to have started as the assessment of tenant right payments, when a tenant quitted an agricultural holding. Such a valuation was known as a tenant right valuation. Tenant right compensation started around 1775 when there was considerable agricultural activity and progress in the country.

1.2 This compensation was paid for on a basis of custom, which developed over the whole country and often varied from county to county and in some cases from estate to estate. The work of assessing the claim was done by farmer valuers, who were well known and respected farmers in the areas in which they lived. These men eventually formed county and area tenant right valuers' associations. Most of these were formed after the passing of the Agricultural Holdings Act 1875, when it became necessary to agree to a uniform basis, in their district, of assessing compensation for the various matters of claim.

1.3 The oldest known valuers' association was founded in the county of Suffolk in 1847. In 1910 a number of these valuers' associations joined together to form the Central Association of Agricultural Valuers, which, today, is the foremost agricultural and tenant right valuers' professional organisation in the country.

1.4 The items to be paid for as tenant right, under custom, were variously: hay, straw, dung, growing crops, acts of husbandry (tillage operations), compensation for feeding stuffs consumed by animals, fallows, offgoing or awaygoing crops, pre-entry (ie entry on to the holding in advance of the date of the tenancy, eg to plant crops), boosey pasture and holdover (ie right to keep a small field for a cow or horses from quitting day of say Candlemas (2 February) until Whitsunday (15 May)).

1.5 With the passing of the Agricultural Holdings Act 1923 and particularly the Agricultural Holdings Act 1948, with their intricacies, the farmer valuers began to die away and the work was undertaken by professional agricultural valuers. However, farmer valuers did not completely disappear until about 45 years ago.

1.6 Prior to this, assessments of the value of land were made for the relief of the poor following the passing of the Poor Relief Act 1601.

1.7 The work of the professionals was further increased during the railway construction era of the 19th century when considerable areas of land were compulsorily acquired and compensation had to be assessed under the provisions of the Land Clauses Consolidation Act 1845.

1.8 The introduction of death duties by Mr David Lloyd George in his budget of 1910 further increased the work of an agricultural valuer as the landed gentry were assessed also on the value of their estates.

1.9 Nevertheless, most of the work of an agricultural valuer, as we know it today, has been extended considerably since the end of the Second World War and is very specialist in nature. Nowadays, it covers not only tenant right valuations, but also rental valuations, agricultural property valuations for many purposes including compulsory acquisition, purchase, sale, inheritance tax and capital gains tax. It also covers compensation claims for easements (gas, water and oil pipelines, electricity lines), compensation for opencast and other mineral extraction, milk quotas, SSSIs, etc.

Definitions

2.1 Termination of tenancies claims

Chapters 1–11 of this book deal in termination of tenancy claims and it is expected that the practitioner has a working knowledge of the Agricultural Holdings Act 1986. Unless otherwise stated, the valuation examples given in Chapters 3–11 are all on the basis prescribed by the 1986 Act and, in particular, the Agriculture (Calculation of Value for Compensation) Regulations 1978 (SI 1978 No 809), the Agriculture (Calculation of Value for Compensation) (Amendment) Regulations 1981 (SI 1981 No 822) and the latest amending Agriculture (Calculation for Compensation (Amendment) Regulations 1983 (SI 1983 No 1475).

2.2 Definitions

In computing the relevant valuations, it is well to be reminded of the following factors which are important.

2.2.1 Pre 1 March 1948 tenancies

Some of these still exist even today, 60 years later. Compensation for tenant right, ie growing crops, harvested crops, root crops, seeds sown, pasture, sod fertility, acclimatisation of hill sheep, etc, will not be paid under the provisions of Part II of schedule 8 of the 1986 Act, unless before the termination of the tenancy, the tenant serves a written notice under para 6 of schedule 12 electing that those provisions shall apply to him. If no such notice is served, compensation for tenant right will be payable under the custom of the district and the provisions of the tenancy agreement.

2.2.2 Consuming value

This is defined in SI 1978 No 809, Part II 8(3) as being the market value for consumption on the holding by agricultural livestock of hay, fodder, crops, straw, roots and other crops or produce of good quality less the manurial value, on a no crop off basis as prescribed by Tables 5(a) and (b) of SI 1983 No 1475.

Local valuers' associations normally fix consuming values for various produce and crops twice a year. These figures reflect the deduction for the manurial values of the crops. The values fixed normally refer to the best quality crops, situated in a good position, etc.

2.2.3 Market value

There is no official definition of market value but this is considered to be what the open market would give for any commodity, without restriction on where it was consumed, eg the market value of a bay of hay is what is considered could be obtained for the same if sold by open auction. Likewise, the market value of an area of swedes is what the open market would give for the same, either for folding off or on harvesting and removal from the holding.

2.2.4 Spring tenancy (SI 1978 No 809, Part II 8/2)

This means a yearly tenancy which commenced between 1 January and 30 June, inclusive.

2.2.5 Autumn tenancy (SI 1978 No 809, Part II 8/2)

This means a yearly tenancy which commenced between 1 September and 31 December inclusive.

2.2.6 Enhancement value (SI 1978 No 809, Part II 8/1(c))

This applies to autumn sown crops where the land is held under a spring tenancy or grass and clover seeds which are sown on the land held under a spring or autumn tenancy where no crop has been taken at the termination of the tenancy.

The enhancement value is an additional amount payable which represents the enhancement of the value of that crop to an incoming tenant, but this shall not exceed the rental value at the termination of the tenancy.

2.3 Tenant right valuations

It is stressed that in all the chapters dealing with tenant right there is no contract between the outgoing tenant and incoming tenant.

The only contract the outgoing tenant has is with the landlord and a tenant right valuation is in theory carried out between the outgoing tenant and the landlord. Normally, however, one valuer acts for the outgoing tenant and another valuer acts for both the landlord and the incoming tenant (if there is one).

2.4 Claims

Under section 83(2) of the Agricultural Holdings Act 1986 landlords' and tenants' claims must be notified before the expiration of two months, ie after the termination of the tenancy. Under section 83(4) of the 1986 Act eight months are allowed for settlement, failing which, the matter shall be determined by arbitration under the Arbitration Act 1996. Note, however, that if the landlord is claiming for general deterioration of the holding, he must give the tenant notice of his intent to claim under section 72 of the Act no later than one month before the termination the tenancy.

Produce

3.1 Produce

Produce comprises harvested hay, straw, silage and roots.

3.2 Hay

Distinguish between seeds and meadow hay. Good meadow hay is fine hay from old pastures and is very suitable for calves and sheep. Seeds hay is from ryegrass, timothy and clover mixtures. Best hay should be leafy, bright, green, smell sweet, be cut young and not be fibrous. Hay suitable for horses has all of these qualities but is seeds hay and free from dust. Avoid fousty, dusty, fibrous, rain washed and overheated hay. Baling too soon and too tightly can spoil hay.

3.2.1 Valuation

3.2.1.1 Conventional bales

Weigh say six bales in a bay, preferably from different layers and get average weight in kgs. Count number of bales, multiply weight by number of bales and divide by 1000 = number of tonnes, eg 920 bales average 20.4 kg in bay:

$$920 \times 20.4/1,000 = 18.768 \text{ tonnes}$$

Volume method

Sometimes it is not possible to count bales. An alternative method is to obtain the cubic content of the stack in m³ and divide by an appropriate density figure:

Very compact:	6–7	m³ per tonne
Medium:	8–9	m³ per tonne
Unsettled:	10–11	m³ per tonne

Practice checks

Small bales of hay often approximate at 50 bales per tonne. Medium and heavy bales often weigh from 18 to 22 kg per bale. Discount any weathered bales on the exposed side of barns and also possibly half or sometimes the whole of the bottom layer as this might be damp. Look out for voids in the stack. The bottom layer has usually more bales than the other layers since they are laid on edge. As a rule a bay 24ft (7.3m) × 15ft (4.5m) × 18ft (5.4m) will contain approximately 20 tonnes.

3.2.1.2 Big bales

Nowadays big balers are used on most farms in preference to conventional balers. Two types are generally used.

(a) Round balers

These are usually of two types each producing a cylindrical bale of about 1.5m wide × up to 1.8m diameter or 1.2m wide × up to 1.5m wide. Note the density of bale made can be controlled and sometimes as a result bales may be very dense and thus heavier. Usually one round bale contains between eight and nine small bales, but this can vary.

(b) Rectangular balers

Two types are in use, namely a low density baler producing a bale about 1.5 × 1.5 × 2.4m in size and a high density baler producing a bale of about 1.2 × 1.3 × 2.4m. Often the weight of hay in big rectangular bales made by a high density baler is anything between 600 and 700 kg.

3.2.1.3 Valuing big bales presents a major problem to an agricultural valuer since sample bales cannot easily be weighed. However, it is essential that say three or four bales are weighed and an average weight obtained. Often this means that they have to be transported to the nearest available weighing facility.

When the weight has been determined, the value can be assessed. This may be on a consuming basis (the local agricultural valuers' association may fix the values) which may need adjusting having regard to:

(a) the quality of the hay
(b) the quantity does not exceed the quantity reasonably required for the system of farming practised
(c) how convenient it is for use
(d) how well it is stacked and protected.

3.2.1.4 Points to note

(a) Big bales are not as easy to handle as small bales. They must be handled mechanically.
(b) Not every farmer has the desire to use big bales and some prefer small ones.
(c) They may have been left out in the fields longer than small bales. The quality of the hay suffers in wet weather conditions especially rectangular bales.
(d) If the bales are out in the open but covered with polythene sheeting, unless placed on concrete they are likely to be damp and there could be considerable wastage.

3.3 Straw

Wheat straw is usually used for bedding only. Spring barley straw is best for feeding. Winter barley straw is of secondary feeding value as it is brittle. Oat straw is also used for feeding but winter oat straw also tends to be brittle. The most valuable straw is, therefore, spring barley straw.

Feeding straw should be bright, not dusty, damp or weathered and the bales light. Wheat and oat straw bales are usually heavier than barley bales. Frequently there are about 60–80 small bales of barley straw to the tonne.

3.3.1 Valuation

3.3.1.1 Conventional bales

The method is as for hay, subject to the following amendments on the density system (where bales cannot be counted).

Heavy:	10–11	m³ per tonne
Medium:	12–13	m³ per tonne
Light:	14–16	m³ per tonne

3.3.1.2 Big bales

Many of the comments made about big hay bales apply to straw and sample bales must be weighed. Often one round bale contains between eight and nine small bales. Often big straw bales are left out in the open but do not suffer, as a rule, if they are wrapped. However, there is considerable wastage and deterioration in quality after a few weeks' exposure to rain and the weather generally where bales are not wrapped. Big bales stacked in a heap tend to become very wet resulting in considerable wastage.

3.3.1.3 Valuation is on the same basis as for hay, using the relevant figures prescribed by the local agricultural valuers' association.

3.4 Silage

3.4.1 Clamp silage

In the absence of market value evidence this should be valued on the basis recommended by the CAAV [Numbered Publication 183] and DEFRA. It should always be analysed and such analysis should be used in assessing both the density and the value of silage. Research by ADAS has shown that there is direct relationship between dry matter percentage and dry matter density of clamp silage. Analysis will give digestibility ('D' values). Values below 50 are poor and above 72 good.

- *Soil contamination* — this is indicated by the ash content which should be under 10%.
- *Fermentable ME* (metabolised energy) — values above 10 megajoules/kilogram corrected dry matter are expected and those below 8 are poor.

- *pH levels* between 3.8 and 4.3 are normal to good while values below 3.8 are poor.
- *Crude protein* (CP) — values in the range 10–16%.
- *Ammonia H* — well fermented value less than 10% of total H.

The bulk density varies considerably with grass clamp silages having densities of between 475 and 1100kg/m³. In a CAVV survey in 2006, 30% of samples had densities between 700 and 900kg/m³. An average density is above 800kg/m³.

Maize silage density is less than grass silage with an average 34% DM of 680kg/m³.

Grass silage has a light green, greenish yellow or greenish brown colour and should smell pleasant, fruity or faintly acid and the texture should be firm. Dark brown or black silage is overheated. Olive green or dark green silage is underheated and has a strong rancid, musty smell. Sometimes secondary fermentation occurs and the silage becomes a brown colour and smells fishy or musty and is rather slimy in texture.

Care should be taken in measuring silage. If the bunker walls are lined with plastic sheeting at the side there is usually a good deal less waste than when unlined. Proper allowance should be made for wastage at the sides, the wedge and also the top where sometimes as much as 150mm (6in) is waste.

Great care should be taken in analysing silage as often three or four cuts are taken and there can be considerable variation in the analysis of the various cuts. Often four samples are taken from four different points on the clamp and all are put in the same bag and analysed as one.

Precision chopped silage in clamps is usually easier to self-feed than single or double chop. Bagged silage is always cut by a mower (not precision chopped) and usually baled direct and then bagged or wrapped.

3.4.2 Big bale silage

This should have sample analysis of at least 5% of the bales. In many cases valuation of bagged and wrapped silage will be found to work within the formula given below, but results provided by high DM percentages should be treated with caution.

If the plastic bags are punctured or badly sealed the silage is of little value and often the whole bag is ruined. The outgoer is responsible for weighing a sufficient number of bales to establish the weight. Market price is the best measure of value as generally there is some sales evidence available.

3.4.3 Maize silage

Maize for silage is a popular crop, grown almost all through the land. It is high in starch, has high digestibility (D value to be 60 to 70) and is low in protein. It can yield from 15 to 25 tonnes per acre. However, it is not a cheap crop to grow and occupies the land for a long period — late April to October.

3.4.4 Valuation

The CAAV recommend that where market value evidence is available this should be taken into account having regard to the valuation basis given in SI 1978 No 708 as amended by SI 1980 No 750, SI 1981 No 822 and SI 1983 No 1475. This is the market value for consumption by agricultural livestock on the holding — produce of good quality less the manurial value thereof calculated on the basis of no crop off, provided that if the value so calculated exceeds the value to an incoming tenant in any case where:

(a) the crops or produce are of inferior quality
(b) the quantity of any kind of crop or produce exceeds the quantity reasonably required for the system of farming practised on the holding or
(c) the crops or produce are left in convenient or proper places on the farm or

the value so calculated shall be reduced so as not to exceed such actual value.

The CAAV recommend that where market value evidence is not available, silage may be valued by reference to the value of the tradeable feeds on the basis of the benefits they offer stock. This approach is based on two baskets of common feeds which are regularly used and traded. One basket contains mainly energy food (feeding barley straw, good hay, barley, wheat, beet pulp and maize gluten). The other basket contains mainly protein foods (Brazilian soya, rape meal, beans, peas and distiller grains).

The method of valuation is to calculate the average energy and protein content of the two baskets. The value of the commodities is published in the weekly reports, as published in *Farmers' Weekly*. Then apply the following formulae:

$$\frac{\text{ME (Average cost of energy feeds} \times 39) - (\text{Average cost of protein feeds} \times 14)}{213}$$

Note: The figures 39, 14 and 213 will remain constant.

$$\frac{\text{CP (Average cost of protein feeds} \times 12.9) - (\text{Average cost of energy feeds} \times 15.6)}{213}$$

Note: The figures 12.9, 15.6 and 213 will remain constant.

These unit values can then be multiplied by the energy and protein contents of the silage (bearing in mind its dry matter) to give a basic value for the silage as follows:

Unit Price ME × ME% × DM%/100 + Unit Price CP × CP% × DM%/100

The resulting value may be adjusted by any of the following factors to reflect a proper value of the silage in question and taking into account transaction evidence of silage value where:

(a) Ammonia levels

Adjustment

Ammonia %	Adjustment	Ammonia %	Adjustment
0–10%	Nil	24%	−23%
11%	−1%	25%	−25%
12%	−2%	26%	−28%
13%	−3%	27%	−31%
14%	−4%	28%	−34%
15%	−5%	29%	−37%
16%	−7%	30%	−40%
17%	−9%	31%	−43%
18%	−11%	32%	−16%
19%	−13%	33%	−49%
20%	−15%	34%	−52%
21%	−17%	35%	−55%
22%	−19%	36%	−58%
23%	−21%	37%	−60%

(b)   Starch content

Formula is:

$$\text{Silage value corrected for starch} = \frac{\text{silage value} \times 100}{(120 - \text{starch \%})}$$

(c) Other relevant factors

The value of the silage should be adjusted having regard to factors such as exceptional and inferior quality, position of clamp on the farm, time of year and whether clamp is open or under cover and whether properly covered.

(d) Valuations under 1986 Act

Deduct for manurial value (Table 5(a) of SI 1978 No 809).

Example valuation

Stage 1

Obtain current prices per tonne for the feeds in the two baskets and calculate the average for each basket. The values used in this example are purely illustrative. They will of course change over time and so will need to be checked for valuations at new dates.

Energy Feeds		Protein Feeds	
Barley Straw	£35.00	Brazilian Soya	£172.00
Hay	£50.00	Rape Meal	£125.00
Barley Grain	£110.00	Beans	£130.00
Wheat Grain	£120.00	Peas	£130.00
Maize Gluten	£135.00	Distillers Grains	£120.00
Beet Pulp	£100.00		
Average	£91.66	Average	£135.40

Stage 2

Assess the cost of a unit of ME in the basket of energy feeds using the ME equation.

(£91.66 {= average cost of energy basket} × 39) − (£135.40 {= average cost of protein basket} × 14)

213

14

$$= \frac{3574.74 - 1895.60}{213} = \frac{1679.14}{213} = \text{£7.88 per unit of ME}$$

NB The figures of 39, 14 and 213 are constants.

Stage 3
Assess the cost of a unit of CP in a basket of protein feeds by using the CP equation.

$$\frac{\text{(£135.40 \{= average cost of protein basket\} × 12.9) − (£91.66 \{= average cost of energy basket\} × 15.6)}}{213}$$

$$= \frac{1746.66 - 1429.89}{213} = \frac{316.77}{213} = \text{£1.48 per unit of CP}$$

NB The figures of 12.9, 15.6 and 213 are constants.

Stage 4
Calculate the feed value of the silage by using the unit values in the equation for energy and protein content.

Assuming the silage has been analysed to show

- ME of 10.6MJ/kg
- DM of 21.0%
- CP of 13.0%

the figures are

$$\frac{7.88 \times 10.6 \text{ ME} \times 21\% \text{ DM}}{100} + \frac{1.48 \times 13\% \text{ CP} \times 21\% \text{ DM}}{100}$$

$$= \text{£17.54} + \text{£4.04} = \text{£21.58 per tonne}$$

Stage 5
Adjust for high ammonia levels using the table at paragraph (a).

Thus, if the silage analysis showed an ammonia content of 12%, then a 2% deduction, 43p should be made which in this case would reduce the value to £21.15 per tonne.

Stage 6

Where maize (or later wholecrop) silage is involved, adjust the value for starch content according to the formula:

$$\text{Silage value corrected for starch} = \frac{\text{silage value} \times 100}{(120 - \text{starch \%})}$$

Stage 7

Value the clamp of silage by making further allowances for any relevant factors.

Stage 8

If required to value on consuming basis (as for the end of a 1986 Act tenancy), deduct for manurial value as follows:

$$\text{UMV Hay} \times \frac{(\text{DM\% of silage})}{85}$$

Stage 9

Again if required by the 1986 Act, reduce the value effecting any inconvenience in location inferior quality or excess quantity of the silage.

3.4.5 Tower silage or haylage

In practice a tower silo of 20ft (6m) diameter has a capacity of 2 tonnes dry matter per foot of depth of silage, a 24ft (7.3m) diameter silo has a capacity of 2.75 tonnes per foot and a 30ft (9.12m) diameter silo about 4.4 tonnes dry matter.

It is strongly recommended that a valuer should study the silo capacity given by the manufacturer.

Advisers recommend that the dry matter capacity (note this is different from the actual capacity — dry matter content of fresh silage is taken at 40%) can be calculated on the following formula:

$$\text{DM capacity (tonnes)} = 6.08 \times d \times 2h \, (17.4 + h)/1{,}000$$

d = diameter in metres
h = settled silage height in metres

Often it works out at 1.55 to 1.77 m^3 per tonne (560 to 65 kg per m^3) at 45 to 50% dry matter (240 kg dry matter per m^3).

In computing the valuation, it is necessary to have the haylage analysed. Difficulty can arise in getting samples from a tower silo except at the bottom and before a valuation is settled, delay may arise in having to wait before further samples can be taken for analysis. Having obtained suitable analysis, it is suggested that the valuation should be done using the same formula as for clamp silage. Normally haylage has a much higher dry matter content than clamp silage and should have a much higher feeding value and thus, a higher value per tonne than silage.

3.5 Harvested roots

The volume is calculated and the following are the amounts in m³ and cu ft per tonne.

	m³ per tonne	cu ft per tonne
Mangolds	1.78	64
Swedes	1.78	64
Potatoes	1.55	55
Fodder Beet	1.78	64

Local valuers' associations, at their price fixing meetings, sometimes fix prices per tonne for these roots (in areas where they are grown) and this is done on the basis of the market value less the manurial value in a no crop off basis in accordance with the SI 1978/809 (as amended).

Growing Crops

4.1 These are valued on the basis of cost of seeds, fertilisers, cultivations, fallows and acts of husbandry plus, in the case of autumn sown crops on a holding held on a spring tenancy, or grass and clover seeds sown in a spring or autumn tenancy where no crop has been taken, an additional amount representing the enhancement value to the incoming tenant of the growing crop not exceeding the annual rental value of the area carrying the crop. Often, the rent paid for the appropriate period of the growth of the crop is added as the enhancement value.

4.2 It is necessary to know the cultivations that are normally carried out. These are usually much less costly on light land than on heavy clay land. Note the action of frost on heavy land ploughed over winter improves the tilth considerably. Also note the increasing use of minimal cultivation or no-ploughing techniques (see below). Full details of all cultivations carried out should be obtained from the farmer. A specimen cultivation procedure on light land for, say, a winter barley crop in Herefordshire may be:

Winter:
- 1 tramline lifting (subsoiling)
- 1 plough and press
- 1 fertiliser spreading
- 1 combination drilling (cultivate and drill)
- 1 pre-emergent spraying for weeds (herbicide)

Spring: 1–3 top dressings
1 spray for mildew
1 spray for weeds (herbicide) may be done in one operation (if not done in the autumn)
1 rolling (possible).

Alternatively, for a crop of winter wheat on heavy land:

Winter: 1 plough
1 roll
2 disc harrows
1 combination drilling (cultivate and drill)
1 pre-emergent spraying for weeds (herbicide)

Spring: 1 roll
1–3 top dressings
2–3 sprayings.

4.3 No ploughing or minimal cultivations

Some farmers, particularly on heavy soils, do not plough, and have found in practice that in using minimal cultivations the top layer of soil has an improved texture and workability and resultant yields are improved. An example of the technique used by an arable farmer on heavy clay in Worcestershire (all winter corn) is:

- 1 spray (if the land is foul) to kill weeds and, if pasture, the grass)
- 1 subsoiling (every three years — but tramline lifting each year)
- 2 or possibly 3 heavy disc harrows
- 1 combination drilling (cultivate and drill)
- 1 harrow
- 2–3 sprayings.

4.4 Valuation

4.4.1 Winter wheat

The valuation is calculated in accordance with the costings issued by the CAAV. An example of a winter wheat crop is as follows:

	£ per acre	£ per ha
Cultivations etc		
1 subsoil (tramline)	19.50	48.20
1 plough	16.55	40.90
Fertiliser spreading	3.15	7.80
1 combination drilling	17.60	43.50
1 spray	2.50	6.20
Seeds (costing £300 per tonne)		
185 kg		55.60
75 kg	22.50	
Fertilisers (0:24:24 @ £180 per tonne)		
308 kg		55.44
125 kg	22.50	
Spray, pre-emergent herbicide	12.00	29.62
Enhancement value	25.00	61.77
	141.30	349.00

4.4.2 Roots (in the ground)

Growing root and green crops (eg kale, stubble turnips) of a kind normally grown on a holding held under an autumn tenancy are valued on the basis of the average market value on the holding of good quality crops less their manurial values as provided by SI tables. In practice, the prices are fixed by the local agricultural valuers' association in the autumn. Example valuations are (as at autumn 2007):

Marrow stem kale	£150 per acre	£370 per ha
Kale (other)	£115 per acre	£284 per ha
Stubble turnips	£70 per acre	£173 per ha

The yield of swedes, mangolds and turnips in the ground is estimated, (this can be checked by cutting and weighing specimen areas) and all are valued in the autumn of 2007 @ £10.15 per tonne.

The above values are for best quality crops and are net of the deduction for manurial values.

4.5 Growing crops — market value basis

4.5.1 Occasionally, crops are required to be paid for on crop value or market value basis, ie what that crop will realise if sold or taken over, as a growing crop on the relevant date. Note, this is most likely to arise under the contracts of sale of farms or form part of agreed surrender terms.

In such valuations it is usual to assess the likely yield, price this out at market value and then deduct the cost of harvesting and also a percentage of risk (usually, dependent on stage of maturity of the crop, 10%–15%) in case it may never be harvested because of weather conditions or losses on lodging (crop being laid). Examples are set out at 4.5.2, 4.5.3 and 4.5.4.

4.5.2 Growing crop of barley

	per acre	per ha
Grain yield 3 tonnes @ £130	390	963
Straw yield 1 tonne @ £35	35	86
	£425	£1,049
less: harvesting, carting and drying	40	98
risk: 10% of gross say	42	104
Crop value	£343	£847

4.5.3 Growing crop of swedes

	per acre	per ha
Estimated net yield 15 tonnes @ £10	150	371
less: harvesting — nil (to be folded off)	–	–
risk: nil (if healthy — but if suffering from disease say 25–50% + can be deducted	–	–
Crop value	£150	£371

Here it is observed that although these values are prescribed by local valuers' associations, it may prove in practice very difficult in some seasons to get these figures on the open market. Also, where folded off, about a quarter of every root is left in the ground as it cannot be eaten. Market value evidence, if available, should be taken into account.

4.5.4 *Growing crop of marrow stem kale*

Note that this crop is very susceptible to frost damage, and it is most unwise to agree to pay anything much for such crop after Christmas as often frosts render the crop valueless. Marrow stem kale has now lost popularity and hybrid kales, which are generally frost resistant and weather durable crops, are of far greater value.

	per acre	per ha
Value (folded off)	£120	£297

Tenant's Pastures

The basis of valuation for tenant's pasture is either of two kinds.

5.1 Cost basis

This basis is to be used where no crop has been removed by mowing or grazing and when sown on land held under a spring or autumn tenancy. The value is the reasonable cost of seeds sown, cultivations, fallows and acts of husbandry performed, also taking into account any expenditure incurred solely for the benefit of the pasture before the removal of any crop in or with which this pasture was sown. Also enhancement value shall be added.

It is very costly to establish leys, as shown in this example.

	per acre	per ha
Seeds (4/5 year ley) say	35.00	86.00
Cultivations namely:		
1 plough and press	19.65	48.55
1 drill	7.10	17.60
1 fertiliser spinner	3.15	7.80
3 rolls (gang)	9.50	23.45
	39.40	97.40
	£74.40	£183.40

Note: Further harrowing or rolling may be necessary

Fertilisers 100 kg 20:10:10	16.00	39.53
247 kg 20:10:10		
Enhancement value say	30.00	74.13
Total cost	£120.40	£297.06

Short term leys (ie of one to three years) mainly of ryegrass and possibly other grasses, cost almost as much to establish, the only exception being that the seeds sown, cost approximately £10 per acre (£25 per ha) less.

If the seeds are undersown (ie drilled in the spring in, say, a growing spring barley crop), the labour costs will be much less.

	per acre	per ha
Seeds say	£20.00	£49.40
Broadcasting seeds (spinner)	£3.15	£7.78
Rolling	£9.50	£23.45
	£32.65	£80.63

However, the practice of undersowing corn with young seeds is now less popular, many farmers preferring to sow them direct, in the autumn, after cereals.

5.2 Face value basis

This basis is applicable where one or more crop has been taken by mowing or grazing. In assessing the value of any particular pasture, the following shall be taken into account.

(a) Present condition. Is it a good quality ley, well established, of a good seeds mixture, clean and free from weeds, especially grass weeds such as couch?

(b) Has it been properly managed since sowing? Leys can be ruined by undergrazing.

(c) Is there water available for stock?

(d) Is the fencing good and stockproof? This is most important

(e) Is it convenient for access and also in relation to the remainder of the farm?

(f) Is it near the end of its useful life and how much longer will it last before it becomes unproductive?

(g) Is the ley dominated by certain grasses and clovers? Cocksfoot is generally a coarse (though productive and drought resistant) grass and if not properly grazed, will dominate a sward. Clover adds considerably to the fertility of the land by nitrogen fixation. However, where there is an excess in a sward, it causes blowing of animals (bloat) which occasionally results in death.

If leys are properly managed, they can improve in quality after they have been established a few years, provided, of course, the seeds mixture is satisfactory and has become well established. Generally speaking, leys do, however, deteriorate considerably towards the end of their life.

Valuing leys accurately at face value is a difficult matter. In the past, valuers have tended to put rather low values on good leys which have considerable unexpired life and productivity. The value can be based on the market value of the pasture, whether for mowing or grazing, taking into account factors (a)–(g) above, with deductions made for the rental value and also any costs that would be incurred by way of application of fertilisers, sprays, pasture topping and, where used for mowing, the harvesting costs. This will produce a value for a season or year but percentage additions should be made for any further years of life the pasture is estimated to have.

An alternative and possibly more accurate basis is to depreciate the cost of establishing the ley, on an annual basis, over its term, less any costs of spraying weeds or pasture topping, but making adjustments for factors (a) to (g) above.

Example valuations

(i)   4 year ley (at start of second year valuation made at Candlemas Day)

	per acre	per ha
Cost of comparable grazing: Summer grazing (7 months)	100.00	247.00
Winter grazing (5 months) Expense incurred supervising stock	20.00	49.00
at away grazing say	5.00	12.00
Gross cost	£125.00	£308.00

	per acre	per ha
Less Deductions		
Rent pa	50.00	123.60
Fertilisers	22.00	54.38
Spraying/pasture topping	5.00	12.36
	77.00	190.34
Net value of ley in		
second year	48.00	118.65
Add for year 3 — 50%	24.00	59.32
Add for year 4 — 25%	12.00	29.66
Value of ley	£84.00	£207.63

(ii) Value of same ley in its last year
(Ley due up for winter
wheat in autumn.
Valuation as at Candlemas)

	per acre	per ha
Cost of comparable grazing		
(7 months) (now a poorer face)	70.00	173.00
Expenses incurred supervising		
stock at away grazing say	5.00	12.36
	£75.00	£185.36
Less Deductions		
Rent 7/12 × £50	29.00	71.65
Fertilisers	22.00	54.36
	51.00	126.01
Net value of ley	£24.00	£59.35

Note: some leys, near the end of their life, may have little or no value,

(iii) Value of same ley (4 year) on depreciating basis (start of second year)

	per acre	per ha
Cost of establishing ley say	100.00	247.10
Less: depreciation for 1 year	25.00	61.75
Value	£75.00	£185.35

This is on the basis of no adjustments being made in respect of (a)-(g) above.

Practice notes

(a) The valuation must ignore any sod value.

(b) If the tenancy agreement schedules land as arable and it is in pasture at the termination date, the pasture is classed as a tenant's pasture and is the subject of a tenant's outgoing claim (unless there is a written agreement that the tenant cannot claim).

(c) If there is no written tenancy agreement in existence and the outgoing tenant can prove, by producing his ingoing valuation inventory, that he paid for any pasture on entry, or that any field was then arable and is now pasture, it can be the subject of a claim by the outgoer.

(d) If the pasture was ploughed up during and after the 1939/45 war, on Agricultural Executive Committee orders, and such written orders can be produced, any pasture subsequently laid down is a tenant's pasture and can be claimed for.

(e) Valuing tenant's pastures is normally a somewhat contentious matter between valuers, who frequently have widely differing opinions on the value of any given pasture.

(f) In making calculations of value for a season or year, do not forget to add for any further unexpired life the pasture will have after this period has expired.

(g) Valuers should carefully note the presence of weeds, eg thistles, docks and dandelions, daisy, etc and weed grasses, eg agrostis and couch grass in pastures. These considerably reduce the value of pastures and are difficult to control.

(h) The availability of a good water supply, shade and shelter and also good stockproof fencing have considerable bearing on value of tenants pastures. Also, very often, stock handling facilities.

Unexhausted Value of Feeding Stuffs, Manures and Lime

6.1 The regulations relating to these are contained in the Agricultural (Calculation of Value for Compensation) Regulations 1978 (SI 1978 No 809) which are known as the principal regulations. The principal regulations regarding this type of compensation have been amended by:

- The Agriculture (Calculation of Value of Compensation) (Amendment) Regulations (SI 1981 No 822). This SI contains the current table for depreciating lime
- The Agriculture (Calculation of Value for Compensation) (Amendment) Regulations (SI 1983 No 1475). These regulations, operative as from 14 November 1983, replaced the previous compensation tables for unexhausted value of phosphate, potash, the value of purchased farmyard manure and the unexhausted manurial values of feeding stuffs consumed. However, note the provisions contained in the principal regulations.

6.2 Unexhausted value of feeding stuffs consumed

6.2.1 Compensation tables are currently found in SI 1983 No 1475. Compensation is payable in respect of corn consumed on the holding (whether produced there or not) or any of the other purchased feeding stuffs listed in the tables that have been purchased and fed to cattle, sheep, horses, pigs and poultry on the holding. This includes purchased hay, straw, etc, and any of the other items, in all 41, which are listed. Note, however, that silage or haylage are not included in this list.

6.2.2 Compensation is payable where no crop has been taken from the resultant farmyard manure or slurry made and on a diminishing scale for crops taken, both after one and two growing seasons. Different rates apply where the crop is fed on the holding and where the product of the manuring, eg hay sold off, is removed from the holding.

6.2.3 All the tables existing in the previous SIs are now superseded by two tables — one for cattle, sheep and pigs (strangely horses are not mentioned here but reference is made to them on p 4 of the SI) and the other for poultry, both on an open slurry basis. This does not necessarily mean that slurry has to be made, as the compensation calculated on the tables is to be adjusted in accordance with Table 6 which deals with whether the manure is kept in a closed slurry pit (ie under slats or in a covered container) or whether it is farmyard manure stored under varying conditions, or whether the feeding stuffs has been fed directly on the land.

6.2.4 When calculating a claim, enquiry has to be made as to how the farmyard manure or slurry is dealt with, as this has considerable bearing on the correct calculation of the compensation payable. It is quite common for some farmers:

(a) to put their stock in strawed yards in early October and cart out the resultant FYM to heap in a field the following spring or summer. It is then spread in the autumn or early winter for the benefit of the following year's crop
(b) in hilly and marginal land areas, in particular, to spread the previous winter's FYM immediately the following spring, on grassland, to produce hay crops
(c) to spread slurry on the land two or three times in the winter and early spring
(d) to leave slurry to solidify in the lagoon, or store, until the summer and then deal with it as (a) above.

6.2.5 Enquiry as to how the manure or slurry is dealt with, will result in the correct information being obtained for the proper calculation of the claim.

6.2.6 Item 41 of Table 5(a) and 21 of Table 5(b) refers to the percentage protein (or albuminoids) in compounded cake, eg 18% albuminoids in 1 tonne of cattle cake consumed, no crop off, is 18 × 20p = £3.60 per tonne consumed.

Specimen claims

(i)   10 tonnes sugar beet nuts fed to beef cattle on slats. No crop off.

10 tonnes × 309p =	30.90	
Conditions ideal — add 30%	£9.27	£40.17

(ii)   20 tonnes milk nuts 16% alb. fed to dairy cattle. After one growing season — hay crop sold off. Farmyard manure made — average conditions.

20 tonnes × (8p × 16%) =		
20 × 128p	25.60	
Add 40% (Table 6)	£10.24	£35.84

(iii) 30 tonnes homegrown seed hay fed to outwintered cattle. After two growing seasons. Product fed on holding. No claim since homegrown hay is not eligible.

Note: If the hay was purchased and brought onto the holding the claim would be:

30 tonnes × 84p =	25.60	
Add 35% (Table 6)	£8.82	£34.02

(iv) 30 tonnes homegrown barley fed to pigs. Farmyard manure made. Ideal conditions. No crop off.

30 tonnes × 204p =	61.20	
Add (Table 6) 50%	£30.60	£91.80

6.3 Unexhausted value of fertilisers applied

6.3.1 Nitrogen (N)

No compensation is payable after one crop has been taken from the land.

6.3.2 Phosphoric Acid (P_2O_5) — Phosphates

Compensation is payable in accordance with Table 2 Part II of SI 1983 No 1475.

Reference should be made to SI 1978 No 809 (pp 6 and 7) which states:

(a) where the phosphatic fertiliser contains less than 1/10 of its total phosphoric acid content in an insoluble form, the total phosphoric acid content shall be treated as soluble

(b) where it is a fertiliser (other than a fertiliser specified and applied as in 2(a); 2(b)(i) or 3(a) in Table 2) containing more than 1/10 of its total in soluble form, the value shall be restricted to the soluble form only. The permanent grassland referred to has to be established for five or more years.

Different rates apply, dependent on the type of fertiliser applied, as detailed in the table and care should be taken to find out exactly what type of phosphate is used. Regarding item 4 of Table 2 calcined calcium aluminium phosphate, to substantiate a claim here, it is necessary to get the soil, where it was applied, analysed. Obviously, if the phosphate level is good, compensation would be payable, by negotiation, of the amount considered fair.

(c) where the phosphate applied is rock phosphate different rates apply where this is soft and these rates have to be adjusted having regard to the amount of the mean annual excess winter rainfall (MAEWR) (different rates apply when the rainfall is above and below 450mm and also when applied to permanent grassland and other crops).

Specimen claims

(i) 10 tonnes Phosphate 23% (20.6% Sol. 2.4% Insol.) applied to pasture. After one growing season.

Calculation is 10 tonnes × (20.6 × 158p)
= 10 × £32.54 = £325.40

(ii) 5 tonnes soft ground rock Phosphate 21% (19.9% Sol. 1.1% Insol.) applied to forage crop in area of 400 mm MAEWR. After two growing seasons.
Claim NIL

(iii) Same quantity and analysis as (ii) applied at same date to permanent pasture in area of less than 450 mm MAEWR. Calculation is:

5 tonnes × (21 × 79p)(2 GS)
5 × £16.59 = £82.95

6.3.3 Potash (K₂O)

Compensation is payable in accordance with the rates shown in Table 3 of SI 1983 No 1475. It is restricted to two growing seasons and then payable only in respect of arable crops (except forage crops), root crops (where the tops are left on the land, eg sugarbeet) and also leys, permanent grassland and forage crops, grazed or the product cut and fed on the holding.

(a) Nothing is payable in respect of potash applied to potatoes, roots, forage crops and applied to leys and permanent grassland where the product (eg hay or silage) is removed from the holding.

(b) The valuer should read paragraph 6(2)(c) of SI 1978 No 809 which, *inter alia*, states that where the holding is mainly horticultural, the value calculated should be in accordance with Item 1 of Table 3 (now SI 1983 No 1475).

Specimen claims

(i) 10 tonnes potash 20% K₂O applied to sugar beet crop (tops part folded off, part ploughed in). After one growing season. Calculation is:

10 tonnes × (20 × 92p)
10 × £18.40 = £184.00

(ii) 10 tonnes potash (20% K₂O) applied to potato field (after one growing season)

No claim.

6.3.4 *Compounds*

Most fertilisers used are compounds of nitrogen, phosphate and potash.

After one crop has been taken, no compensation is payable in respect of nitrogen but the compensation payable in respect of phosphates and potash shall be in accordance with the tables in SI 1983 No 1475.

Specimen claim

(i)   10 tonnes compound 20%.10% (1% Insol.).10% (the analysis is always shown in the order Nitrogen:Phosphate:Potash) applied to permanent grazing meadows. Claim after one growing season is:

Nitrogen	Nil	
P_2O_5 10 tonnes $\times$ (10 $\times$ 158p) =	£158	
K_2O 10 tonnes $\times$ (10 $\times$ 92p) =	£92	
Total		£250

(ii) 10 tonnes maincrop potato compound 13.13(1.2 Insol).20 applied to main-crop potatoes — after two growing seasons. MAEWR 460 mm.

Claim is:		
Nitrogen	Nil	
P_2O_5 10 tonnes $\times$ (13 $\times$ 79p) =	£102.70	
K_2O_5	Nil	
Total		£102.70

6.4 Farmyard manure

Compensation for home produced farmyard manure is calculated through the feeding stuffs consumed and nothing, otherwise, is paid, except for the labour involved. Note maximum compensation is for 50 tonnes/ha [20t/a] cattle, horse and pig manure; 18 tonnes/ha [7.28t/a] deep litter poultry manure and 12.5 tonnes/ha [5t/a] broiler poultry manure.

6.4.1 If farmyard manure (this includes poultry and horse manure) is brought onto the holding and no crop has been taken since the manure was applied, the value shall be for the cost of delivery and application where no payment was made for the manure. In this case the value in subsequent years, after the first growing season, is half, and after the second growing season, is one quarter, of the cost of delivery and application (SI 1978 No 809). No compensation is payable in respect of purchased manure in the last year of tenancy, after the last crop was removed, unless the landlord gave written consent for such application.

6.4.2 Where payment is made for the manure, the claim shall be for the cost of the manure, its delivery and application where no crop is taken, and after one and two growing seasons, this shall be reduced to one half and one quarter respectively, provided the value of the manure specified in Table 4 of SI 1983 No 1475 shall not exceed the figures given in that table.

Specimen claims

(a)　100 tonnes purchased broiler manure, after two growing seasons.

Cost including spreading	£1,560
Claim — quarter cost =	£390

(b)　80 tonnes pig farmyard manure (free for fetching) after one growing season.

Cost of collection and spreading	£760
Value after 1 season, half =	£380

6.5　Unexhausted value of lime applied

Compensation provisions here are contained in SI 1981 No 822 p 3. The calculation appears somewhat complicated, but is not so, providing the following are ascertained:

(a)　the amount of the mean annual excess winter rainfall (definition given in SI 1981 No 822)
(b)　the amount of nitrogen applied annually to the land concerned. If the land is in permanent pasture and long term leys or in arable rotation (with short term leys) and more than 250kg of nitrogen per ha is applied annually, the cost of the lime applied (this includes cost of delivery and application) can be depreciated, dependent on mean annual excess winter rainfall from four years

(where the rainfall exceeds 500 mm MAEWR) up to eight years where it is less than 250m.

If nitrogen applied is up to 250kg per ha annually, and the land concerned comprises permanent pasture or long term leys, the cost of the lime applied is depreciated over nine years where the MAEWR is less than 250 mm and over five years where it is more than 500mm.

In cases when the rainfall is between 250 and 500mm the depreciation is over seven years.

Note 1: For period of one year following the application of the lime the value shall be its cost applied.

Note 2:

(a) The cost shall not be regarded as reasonable to the extent that it exceeds the higher of:

   (i) the cost, at time of application, of the limestone or chalk applied (whichever is cheaper) which would have been used in the application to the land of calcium oxide at a rate 7.33 tonnes per ha and

   (ii) the cost of lime recommended in scientific advices relating to the condition of the soil.

(b) Cost includes delivery and application.

Specimen claims

(a) Lime applied to long term pasture after two growing seasons. Cost including delivery and spreading was £1,050.

450 mm MAEWR150 kg N per ha applied annually. Claim will be:

Cost depreciated for 2 years = 2/7 off cost £1050 = £750.

(b) Lime applied to field in regular arable cropping after four growing seasons. Cost including delivery and spreading was £500. Average of 350kg N per ha applied annually.

210 mm MAEWR.

Claim will be:

Cost depreciated for 8 years = half off cost £500 = £250.

Sod Fertility Claims

7.1 This type of claim was introduced by SI 1978 No 809 and this has since been amended by SI 1980 No 751 and later by SI 1983 No 1475.

7.2 This claim is intended to compensate an outgoing tenant for increased fertility resulting from leaving more pasture (and thus with more inherent fertility) than he[1] is required to, under the terms of the tenancy agreement.

7.3 For a claim to succeed it is important to have detailed records available.

7.4 A stipulation in SI 1983 No 1475 states that the holding must be in an area 'where arable crops can be grown in an unbroken series of not less than 6 years and it is reasonable that they should be grown on the holding, of part thereof'.

7.5 The calculation is very complicated and a claim will not succeed unless very detailed and reliable records can be produced, for several past years, of the cropping of the various fields forming the holding.

7.6 Explanation of matters referred to in paragraph 12 of SI 1978 No 809 are set out below.

(a) Accepted proportion of leys

This is the area specified in the tenancy agreement of leys or new seeds to be left at the tenancy termination. If there is no agreement or it is silent on the matter it is the area representing the proportion which the total area of leys on the holding would, taking into account the capability of the holding, be expected to bear to the area of the holding, excluding permanent pasture.

(b) Excess area of leys

This is the average of the total area of leys at the termination date, and one and two years prior to that date less the accepted area required for the fertility of the holdings. The surplus is the excess area which may be considered for compensation.

(c) Qualifying ley

This is an established ley (three or more years old) or what was a former established ley, that is one which was three years old or older before ploughing or destroying.

Note: Permanent pasture is excluded, as are leys laid down at the expense of the landlord, without payment, by the tenant or any previous tenant.

(d) Calculation of compensation

The residual value of the sod of the excess qualifying leys is calculated, but subject to the amendments contained in paragraph 3(2)(d)(i) and (ii) of SI 1983 No 1475 as follows:

(i) in respect of continuously (three or more growing seasons) maintained leys, £24 per ha if the crop has been cut and removed in the last growing season or £40 per ha if grazed only in the last season

(ii) regarding continuously maintained leys, the value shall be increased by £8 per ha for each additional growing season over three, but subject to a maximum of £48 per ha where the crop was cut and removed or £64 per ha where grazed only in last season

(iii) in respect of any former ley where the first crop which has been sown in the last growing season before termination of tenancy has not been removed from the ground, the value shall be the value specified in (i) and (ii) above according to the period for which the ley had been established before ploughing up or destroying and to whether the herbage was cut and removed, or grazed only, in the last season before ploughing etc

(iv) in any former ley to which (iii) does not apply

(i)(aa) if only one arable crop was taken following ploughing etc, value is two-thirds of that specified in (i) and (ii) above

(bb) if two arable crops taken, value is one-third according in each case to the period for which the ley was established and to whether the ley was cut and removed or grazed in last season before ploughing

(ii) if more than two arable crops taken since ploughing — value is nil.

Where a tenant is entitled to compensation for both tenant's pasture (under paragraph 10 SI 1978 No 809) and sod value under (d)(i) and if applicable (d)(ii) above the aggregate of the respective values per hectare, taken together, shall not exceed £164 per ha.

An example of a claim on a 317 acre (128.3 ha) holding is as follows.

Sod value claims

Arable acreage on schedule of agreement: 239.522 acres (96.97 ha)
Area to be left as Leys on Agreement: 2/5 = 38.78 ha

OS No	Leys Acre	Ha	Description	Year laid down	Last year
Pt 54	6.00	2.42	Long Ley	2000	Grazed
54	16.569	6.70	Long Ley	2000	Grazed
53	17.143	6.94	Long Ley	2000	Grazed
619	11.691	4.73	Long Ley	2000	Mown
651	17.374	7.03	Long Ley	2000	Mown
654	14.043	5.68	Long Ley	8 years ago	Grazed
58	9.370	3.79	Long Ley	8 years ago	Grazed
Pt 149	12.503	5.06	Long Ley	8 years ago	Grazed
221	8.649	3.50	Long Ley	8 years ago	Grazed

Cropping after leys

220	18.341	7.42	Barley 2007	Wheat 2006
				Previous 4 years Ley
				Last mown 2005
617	9.368	3.79	Roots 2007	Wheat 2006
				Ley 2000–2005
				Mown 2005
604	7.263	2.94	Barley 2007	After 4 years Ley sown 2004
				Mown 2005
652	8.202	3.32	Barley 2007	Mown 2006
653	4.985	2.01	Barley 2007	Barley 2006
				After 4 years Ley sown 2002
				Grazed 2005
656	15.094	6.11	Wheat 2007	Pasture for 10 years
				Grazed 2005

Sequence of leys

OS No	Acres	Ha	2000	2001	2002	2003	2004	2005	2006	2007
Pt 54	6	2.42	L	L	L	L	L	L	L	L(G)
Pt 54	16.569	6.70	L	L	L	L	L	L	L	L(G)
53	17.143	6.94	L	L	L	L	L	L	L	L(G)
619	11.691	4.73	L	L	L	L	L	L	L	L(M)
651	14.043	5.68	L	L	L	L	L	L	L	L(M)
654	14.043	5.68	L	L	L	L	L	L	L	L(G)
58	9.370	3.79	L	L	L	L	L	L	L	L(G)
Pt 149	12.503	5.06	L	L	L	L	L	L	L	L(G)
221	8.649	3.50	L	L	L	L	L	L	L	L(G)
220	18.341	7.42			L	L	L	L(M)	W	B
617	9.368	3.79	L	L	L	L	L	L(M)	W	R
604	7.263	2.94				L	L	L(M)	L	B
652	8.202	3.32		L	L	L	L	L	L(M)	B
653	4.985	2.01			L	L	L	L(B)	B	B
656	15.094	6.11	L	L	L	L	L	L	L(G)	W

All qualifying leys as are continuously maintained leys or former leys

L	= Ley	G	= Grazed	R	= Roots
M	= Mown	B	= Barley	W	= Wheat

Eligible leys in excess of area reasonably required for fertility

1. Area of qualifying leys in each of last three years sown all as spring sown

OS No	2007	2006	2005
Pt 54	2.42	2.42	2.42
Pt 54	6.70	6.70	6.70
53	6.94	6.94	6.94
619	4.73	4.73	4.73
651	7.03	7.03	7.03
654	5.68	5.68	5.68
58	3.79	3.79	3.79
Pt 149	5.06	5.06	5.06
221	3.50	3.50	3.50
220			7.42
617			3.79
604		2.94	2.94
652		3.32	3.32
653			2.01
656		6.11	6.11
	45.85	58.22	71.44

Mean 58.50 ha

Less
2. Total area of leys to be left sown, reasonably required for fertility of holding is 2/5 of arable area of 96.9 ha = 38.78 ha.

3. *Equals*
    58.50 ha – 38.78 ha = 19.72 ha Compensationable

Valuation summary
GS = No of growing seasons including year sown
G = Grazed   M = Mown

OS No	Ha	GS	G/M	Years cropped after grass	SFV per ha	
Pt 54	2.42	3	G		£40.00	
Pt 54	6.70	3	G		40.00	
53	6.94	3	G		40.00	
619	4.73	4	M	(24 + 8)	32.00	
651	7.03	4	M	(24 + 8)	32.00	
654	5.68	8	G		64.00	(max)
58	3.79	8	G		64.00	(max)

OS No		Ha	GS	G/M	Years cropped after grass	SFV per ha		
Pt	149	5.06	8	G			64.00	(max)
	221	3.50	8	G			64.00	(max)
	220	7.42	6	M	2	(24 × 1/3)	8.00	
	617	3.79	6	M	2	(24 × 1/3)	8.00	
	604	2.94	5	M	2	(24 × 1/3)	8.00	
	652	3.32	7	M	1	(24 × 2/3)	16.00	
	653	2.01	6	G	2	(24 × 1/3)	8.00	
	656	6.11	8	G	1	(24 × 2/3)	16.00	

Fields selected for maximum compensation

654, 58, Pt 149, 221	18.03	ha @ £64	£1,153.92
Pt54	1.69	ha @ £40	67.60
Total	19.72	ha	
Total compensation			£1,221.52

Note: The SFV per ha is prescribed in SI 1978 No 809, as amended by SI 1980 No 751, as further amended by SI 1983 No 1475.

---

1   In this text he/she are used interchangeably and refer to both male and female.

Tenant's Improvements

8.1 This chapter deals with long term improvements, eg erection of buildings, reclamation of land, making of roads, provision of sheep-dipping accommodation for which the landlord's consent or approval of the Agricultural Lands Tribunal is required.

Distinction should be made between old long term improvements — those carried out *before* 1 March 1948 and those called new long term improvements — those carried out *after* 1 March 1948.

8.2 Old long term improvements (schedule 9 Part I, 1986 Act) — these are major improvement works, such as building erection and are still frequently encountered despite being over 60 years old. If the consent given was an open one and did not specify agreed terms of compensation or is under custom, then paragraph 2 of schedule 9 to the Agricultural Holdings Act 1986 provides that the compensation payable shall be 'an amount equal to the increase attributable to the improvement in the value of the Agricultural Holding as a holding, having regard to the character and situation of the holding and the average requirements of tenants reasonably skilled in husbandry'.

8.3 Long term new improvements (schedule 7 Parts I and II, 1986 Act) — these again are major improvements, broadly in line with old long term improvements, but not entirely so.

8.3.1 These improvements again require the landlord's and/or the Agricultural Land Tribunal's consent. Many landlords and their agents give consent over a set write-off period, often over 10 or 15 years, (but in the case of specialised buildings such as pig or dairy units, sometimes less) with the net cost (after any grant) being written down to nil. Any write-off should be down to say £1 — if only to recognise the tenant's ownership of the improvement.

8.3.2 It is considered by many valuers that 10–15 years' write-off period is far too low for substantial buildings such as a covered yard or Dutch barn which can reasonably be expected, provided it is well built, designed and properly maintained to last 50 years or so.

In the Agricultural Land Tribunal case, *Barton* v *The Lincolnshire Trust for Nature Conservation* (1997), ruled that, since the improvement concerned, in this case, a 6,000,000 gallon reservoir for crop irrigation, had a substantial value extending far beyond the write-off period and overrode the landlord's requirements, that there should be a write-off and substituted the statutory compensation under section 66 of the Act.

8.3.3 If the landlord gives an open unrestricted consent (which, incidentally, can also be given after the improvement has been effected) compensation provisions on the termination of the tenancy are governed by section 66 of the 1986 Act, and the amount of compensation shall be an amount equal to the increase attributable to the improvement in the value of the holding as a holding, ie a let farm, and regard has to be had to the character and situation of the holding and the average requirements of tenants reasonably skilled in husbandry. This infers that the improvement must be suitable for the holding, eg it could be inappropriate to have an expensive dairy set up on a hill farm, or possibly an expensive grain store on a wet farm which is more suitable for grazing than for arable production. The usual method of valuation is to assess the increase in the rental value of the holding arising from the improvement and capitalising this at a suitable rate of interest for its remaining estimated life and thus produce the increased value of the holding. Regard must be had of the likely life of the improvement, eg an electrical wiring installation will probably need renewing in 15 years' time, while a Dutch Barn may last 50 years. A number of examples are set out below.

(a) Covered cattle yard 90 ft × 50 ft (27.36m ×15.2m) erected by tenant four years ago. Cost £30,000. Farm 150 acres in area and short of modern buildings.

This is a good substantial modern building that is needed on the farm and can be put to almost any use, and is expected to last 30 years. It is considered that the erection of this building has increased the rental value of the holding by £2,000 pa. The claim will therefore be:

Increased rental value (net)	£2,000
YP 26 years @ 9%	9.93
Increase in value of holding	£19,860

(b) A tenant has reclaimed an area of 20 acres rough land, including providing certain drainage and fencing works. It is now good pasture with a rental value of £40 per acre which is expected to continue (subject to inflation or deflation) in perpetuity. Its unimproved value would have been £10 per acre. The claim would be calculated as follows:

Current rental value 20 acres @ £40	£800	
Less: Unimproved Rental Value	£200	
	£600	
YP 20 years @ 9%	9.13	
	£5,478	say £5,500

8.4 Note: The relevant section of the 1986 Act should be studied carefully. Sometimes, tenants have carried out long term improvements without landlord's consent and cannot, therefore, obtain compensation. However, their position may not be totally lost as, by giving the requisite notice, the items may be treated as tenant's fixtures, and the improvement can be removed or be taken over by the landlord, should he elect to do so.

8.5 Note that application of purchased manures, liming and consumption on the holding of corn (whether produced there or not) and of cake or other feeding stuff not produced on the holding are improvements and not tenant right. Landlord's consent for their use, however, is not needed.

Tenant's Fixtures

9.1 Section 10 of the Agricultural Holdings Act 1986 gives the tenant the right to remove buildings, engines, machinery, fencing or other fixtures of whatever description affixed to an agricultural holding by the tenant, provided the fixture is not affixed in pursuance of some obligation to do so. The tenant may remove the fixture at any time during the tenancy or before the expiry of two months from the termination of the tenancy, providing he has paid all rent owing and satisfied all his tenant's obligations and also has, at least one month before the exercise of the right and the termination of the tenancy, served on the landlord a written notice of his intention to remove the fixture or building. The landlord can serve on the tenant a written counter-notice before the expiration of the notice electing to purchase the fixture etc, and the measure of compensation payable by the landlord is the 'fair value of the fixture or building to an incoming Tenant'. Any dispute as to the amount of compensation shall be determined by arbitration.

9.2 This simply means that the compensation is what an incomer would be expected to give for the fixture or what it is worth to him.

Factors to take into account in assessing this compensation are:

(i) the current net cost of providing the fixture
(ii) the existing condition of the fixture, taking into account the cost of any repairs necessary

(iii) whether the fixture is a modern one or not — perhaps it may be outdated eg an abreast milking parlour may be considered outdated, or a building erected may have too low a headroom

(iv) the expectation of the fixture's remaining useful life

(v) is the fixture of value to an incoming tenant or not? For example, a landlord may have elected to take over a piggery erected by the outgoing tenant, but the incoming tenant does not intend keeping pigs and to the incomer it may only be of use for his ewe flock at lambing time. It should, however, be borne in mind that in this example, the valuation is between the outgoing tenant and landlord, and since the landlord has elected to take to the fixture, it presumably has value to the incomer. There may be other factors to be considered, but the valuation of fixtures, as in the case of tenant's improvements, is difficult and often arguable.

Examples

(*Note*: In each case below counter-notices are deemed to have been served to purchase the fixtures, all of which will be of value to the incoming tenant.)

(i) The outgoing tenant erected, 10 years ago, a 10/20 herringbone parlour and adjacent dairy. He provided all the milking equipment including 4,000 litre bulk milk tank milking equipment, etc. The total cost of the work was £20,000 for the buildings and £15,000 for the equipment. Repairs estimated to cost £2,000 are necessary to the buildings and £2,000 for the equipment. The buildings are estimated to last a further 10 years and the equipment five years, when both will need replacing at current estimated net cost, respectively, of £25,000 and £25,000. It is suggested that the claim be dealt with as follows:

Building			
Cost (net) of replacement		£25,000	
Less: 1/3 life expired	£8,333		
Repairs	£2,000	£10,333	
			£14,667
Equipment			
Cost of replacement		£25,000	
Less: half life expired	£12,500		
Repairs	£2,000	£14,500	
			£10,500
Net Value			£25,167

(ii) Pig netting fencing of 300m run and one 4.5m wide gate erected in OS No 12 five years ago. Repairs estimated to cost £50 are necessary. The whole is expected to have a life of 15 years from date of erection. Current net cost of the work is estimated to be £940.

Estimated replacement costs		£940
Less: 5/15 life expired	£313	
Repairs	£50	£363
		£577

(iii) A sheep dip, foot bath and handling pens constructed five years ago. The replacement cost today would be £10,000. Repairs estimated to cost £1,000 are necessary. The set-up has an estimated unexpired life of five years and will be of use to the incoming tenant. The claim would be:

Replacement cost		£10,000
Less: Repairs	£1,000	
Half life expired	£5,000	£6,000
		£4,000

Offgoing Crops (or Away Going Crop)

10.1 Certain old tenancy agreements sometimes provide for an acreage of offgoing crop of winter wheat. These tenancies were always Candlemas (2 February) or Lady Day (25 March) and the principle was that the outgoing tenant should retain the crop he planted and that he should be allowed back on the farm to harvest his crop, despite the tenancy having terminated. This proved inconvenient and it became standard practice to value the crop and commute the profit into a cash payment. In effect, the offgoing crop right is the profit on a crop of wheat, less risk (it may become diseased etc). At a meeting of a local Agricultural valuers' association, held in Hereford in 1957, the following matters and basis of valuation were agreed:

(i) that where the right is established and unless the agreement states to the contrary, the offgoing crop right should be 25% of the arable area

(ii) unless the agreement decrees otherwise, the crop must be wheat

(iii) the crop must be planted by Candlemas Day (2 February)

(iv) the valuation is the estimated value of the crop (the straw is left free to the new tenant) deducting harvesting costs and a percentage for risk.

Example valuation of offgoing crop right (per acre)

Yield: 3 tonnes @ £160   £480

Less: Spraying	£40	
Harvesting	£40	
Risk 25% say	£120	
	£200	
Profit ie value of right		£280 (£691 per ha)

Example of Claim at the Termination of a Tenancy (2 February 2008)

AGRICULTURAL HOLDINGS ACT 1986

Re: The Holding known as
           WEST FARM, METCHINGFIELD, HEREFORD
To: THE WALFORD ESTATE COMPANY (Landlords)
           ESTATE OFFICE, METCHINGFIELD, HEREFORD.

I HEREBY GIVE YOU NOTICE pursuant to Section 83(2) of the above Act of my intention to make against you certain claims arising out of the termination of the tenancy of the above holding, the nature of which claims is set out in the Schedule hereto.

I also hereby give you notice, under Section 13 and Schedule 1 of the Agriculture Act 1986, that it is my intention to make against you the claim specified in the Schedule hereto, being a claim for compensation in respect of Milk Quota, which claim arises under para 1 of Schedule 1 of the Said Act on the termination of my tenancy of the said holding.

THE SCHEDULE

Claims made under Clause VII of the Tenancy Agreement dated 4 April 1972 and the Surrender Agreement dated 20 September 2007 and principally in accordance with SI 1978 No 809, SI 1981 No 822 and SI 1983 No 1475

1. **PRODUCE**
   - (a) **BARLEY STRAW**
     Long Barn
     84.6 tonnes 2007 Baled Straw @ £25   £2,115.00
     Home Barn
     25 Big Bales 2006 Straw Av weight
     230kg 5.75 tonnes @ £25   £143.75

(b) **HAY**
Long Barn
25.4 tonnes 2007 Seeds Hay @ £80   £2,032.00
Covered Yard
42 Big Bales 2007 Meadow Hay
Av weight 510kg–21.42 tones @ £70   £1,499.40

(c) **SILAGE**
Clamp at Homestead
Measurements 25m × 13 m × 2m
Value of Silage per tonne is £20.96
422.5 tonnes (separate calculation attached) £8,855.60
               £14,645.75

2. GROWING CROPS

(a) **OS Pt 602 8.79 ha — Sonja Winter Barley**
8.5 ha Planted

1 Subsoiling (tramline)	£27.85		
1 Plough (5 furrow) and press	£36.95		
1 Disc harrow	£18.80		
1 Combination drill	£17.60		
1 Fertiliser spinner	£7.30		
1 P/E Spray	£7.30		
8.5 ha @	£115.80	£984.30	

Seed
30 × 50kg @ £20 per 50 kg   £600.00

Fertilisers
42 × 50kg 9:25:25
(Ex MSF A/C 11) @ £160 per tonne £336.00

Sprays
(Ex MSF A/C 12)   £45.00

Enhancement Value
£60 per ha @ 8.5   £510.00 £2,475.30

(b)

OS No (pt) 652	3.32 ha
OS No 604	2.97 ha
OS No 220	7.42 ha
OS No 218	6.87 ha
	20.58 ha

AVALON WINTER WHEAT 20 ha Planted

1 Subsoiling (tramline)	£27.85	
1 Plough and press	£36.95	
2 Disc harrows	£37.60	
1 Combination drill	£17.60	
1 P/E Spray	£7.30	
20 ha @	£127.30	£2,546.00

Seed
74 × 50kg @ £20 per 50 kg
(ex MSF A/C 10) £1,480.00

Fertilisers
100 × 50kg 9:25:25 @ £160
per tonne (ex MSF A/C 11) £800.00

Sprays (ex MSFA/C 12) £105.88

Enhancement Value
£60 per ha @ 20 ha £1,200.00 £6,131.88

(c) **OS No 618 1 ha SWEDES**

(Planted 16 June 2007)

1 Plough	£36.95	
1 Disc harrow	£18.80	
1 Roll	£13.45	
1 Harrow	£18.80	
1 Drill	£31.25	
1 Fertiliser spinner	£9.85	
1 ha @	£129.10	£129.10

Seed
(Ex MSF A/C 14) £30.00

Fertilisers
13 × 50 kg Compound
@ £160 per tonne £104.00 £263.10

£8,870.28

3. **CULTIVATIONS**

OS No 653	2.01 ha	
OS No 616	4.56 ha	
OS No 832	5.25 ha	
OS No 617	2.43 ha	
Ploughed 14 ha @ £36.95		£517.30

4. **LABOUR TO FYM**
   (a) **Heap in OS 655**
       22m × 15m × 2/3
       220m^3 @ £2.00   £440.00

   (b) **Heap in OS 52**
       11 m × 15m × 2/3m
       110m^3 £2.00   £222.00   £662.00

5. **TENANTS PASTURES** (All scheduled Arable in Tenancy Agreement)
   (a) **OS No 54 9.10 ha**
       4 year ley in last year
       9 ha @ £50   £450.00

   (b) **OS No 53 6.94 ha**
       4 year ley sown Aug 2007
       6.75 ha @ £250   £1,687.50

   (c) **OS No 651 7.02 ha**
       OS No 654 6.68 ha
       13.70 ha
       Long term ley sown 2004
       13.5 ha £150   £2,025.00

   (d) **OS No (pt) 145 5.06 ha**
       4 year ley sown after winter barley
       — Aug 2006, 5 ha @ £220   £1,000.00

   (e) **OS No 58**   3.79 ha
       OS No 221   3.50 ha
       OS No 619   4.72 ha
                                 ──────────
                                 12.01 ha

       Permanent leys laid down in 1988
       11.5 ha @ £75   £862.50

   (f) **OS 280 6.50 ha Young Seeds 3 Year Ley**
       Sown August 2007 (not grazed)
       1 Subsoiling (tramline)   £28.75
       1 Plough and press   £36.95
       1 Roterra   £26.95
       1 Drill (broadcast)   £7.80
       1 Fertiliser spinner   £11.75
       1 Roll (3 gang)   £10.10
       6 ha @   £122.30   £733.80

Seeds
(Ex MSF A/C 15) £360.00

Fertilisers
30 × 50 kg 20:10:10 @ £150 per tonne £225.00

Enhancement Value
6 ha @ £70 £420.00

| | £1,738.80 | £7,763.80 |

6. **SOD VALUES** (As separate calculation) £467.84

7. **RESIDUAL VALUE OF FEEDING STUFFS CONSUMED** £987.00

8. **UNEXHAUSTED MANURIAL VALUES** £1,262.87

9. **RESIDUE OF LIME APPLIED** £421.61

10. **TENANT'S IMPROVEMENTS**
As already agreed £3,260.00

11. **MILK QUOTA**
Allocated quota in excess of standard quota;
Tenant's fraction of standard quota;
Purchased quota of which tenant has borne
entire cost.
As already agreed £15,051.00

SUMMARY

1.	Produce	£14,645.35
2.	Growing Crops	£8,870.28
3.	Cultivations	£517.30
4.	Labour to FYM	£662.00
5.	Tenant's Pasture	£7,763.80
6.	Sod Values	£467.84
7.	RVFS	£987.00
8.	UMVs	£1,262.87
9.	Residue of Lime	£421.61
10.	Tenant's improvements	£3,260.00
11.	Milk quota	£15,051.00
		£53,909.05

Dated this 30th day of January 2008

. . . .
WILLIAMS, DAVIES AND REES,
Agricultural Valuers,
Dolanog,
Welshpool,
Powys.

Authorised Valuers Agents on Behalf of the Outgoing Tenant
Mr J Yeoman — Farmer

E&OE

Dilapidations

11.1 Liability

The landlord of an agricultural holding, can, under the provisions of section 71(1) of the Agricultural Holdings Act 1986, on the tenant quitting the holding, on the termination of the tenancy recover compensation in respect of dilapidations, or deterioration or damage to any part of the holding by the nonfulfillment by the tenant of his responsibility to farm the holding in accordance with the rules of good husbandry. These rules are laid down in section 11(1)–(3) of the Agriculture Act 1947 and briefly cover proper maintenance of pasture, keeping the land clean and in a good state of cultivation, keeping the farm properly stocked, maintaining efficient management of livestock, maintaining crops and livestock free from disease, etc, protection and preservation of harvested crops and carrying out necessary maintenance and repair work.

The amount of the compensation recoverable shall be the cost, at the date of quitting, of making good the dilapidation, deterioration or damage (section 71(2)).

However, the landlord can, instead of claiming under section 71(1), claim under the terms of a tenancy agreement (section 71(3)), which is the normal basis under which most dilapidation claims are made. Where a claim is made under the provisions of the tenancy agreement, section 71(4)(b) states that compensation shall not be claimed in respect of any one holding, under both such tenancy agreement and section 71(1).

Section 71(5) provides that the compensation recoverable under section 71(1) and 71(3) shall not exceed the amount (if any) by which the value of the landlord's reversion in the holding is diminished owing to the dilapidation, deterioration or damage in question.

This latter overriding provision, which was introduced in 1984, has resulted in some dilapidation claims becoming very contentious. The reason for this is that when an agricultural tenancy, protected by the 1986 Act terminates, it can be justifiably argued that there is no loss to the landlord since the value of the property, now with vacant possession, is immediately doubled. Similarly, if the holding is re-let, the rental obtainable on a letting under the Agricultural Tenancies Act 1995 is again much enhanced on the previous rent paid. One day the courts will decide this issue.

11.2 Note the difference between dilapidations and general deterioration. Section 72 of the 1986 Act gives the landlord an additional remedy in so far as a specific claim for dilapidations does not effectively give him adequate compensation for his loss. Examples of the deterioration of a holding are allowing the farm to be severely neglected in almost every respect, including severe loss of soil fertility or serious infestation of diseases, weeds, etc, possibly resulting in lower yields for some period in the future. A claim for deterioration will not be effective unless at least a month's written notice of intention of claim has been given before the termination of the tenancy.

11.3 Any claim for dilapidations arising at the termination of a tenancy must be served within two months of the termination date. It is sufficient to give a notice specifying the nature of the claim and also under what term of the tenancy agreement, rules of good husbandry, model clauses, etc, it arises. In practice a claim is lodged in detail and this also specifies the amount of each item of claim. In preparing each claim it is sound practice to specify the item of dilapidation with clear reference to where it exists, eg in which OS number or on what boundary between one OS number and another. Each item of claim should be numbered (this is useful in subsequent negotiations), the basis of computing the claim should be shown together with the pricing mechanism, eg two worker days @ £88.85 or 105m of overgrown hedge to cut and lay, clear up brash and burn, £12.02 per m (114.5 yds @ £11.00).

11.4 The claim should be set out properly for ease of reference. Some valuers, where large claims are involved, use quantity paper (which has several columns ideal for number of worker days, tractor and implement days, spot items, etc). However, it is preferred that the claim should be kept as simple and clear as possible.

The item of dilapidation should always be first specified, followed by the remedy needed and the cost thereof, eg

No	Ref	Item	Rate	Total

OS No 6531 — 6.8 ha stubble
    1   Foul with Couch, Spray with Roundup 2008
             6.5 ha
    1a   Spray with half dose — do in 2009

OS No 6531/6888 — Very high overgrown hedge
    2   Cut and lay 224 m

OS No 6531/6888 — Wide blocked ditch
    3   Dig out and leave clean 200m

OS No 7331 — 10.85 ha pasture
    4   Pond in NE corner silted up. Clean out.
             10 hrs drag line

11.5 Costing

Costing claims accurately can be very difficult. There is such a variation in different parts of the country of relevant matters such as availability of labour, skill, types of soil, topography, location, etc. The annual costings produced in October of each year by the Central Association of Agricultural Valuers is a most valuable guide and can be adapted for many of the items that are likely to arise in any claim, eg if field is foul with couch the cost can be calculated on the basis of cost of spraying say £2.50 and cost of Roundup say £15.00 = total of £17.50 per acre or £43.24 per ha. If a further application at half rate is needed the following year the claim will be again £2.50 plus £7.50 for the spray — a total of £10.00 (£24.71 per ha).

11.6 Most of the work of remedying dilapidations is done by the farmer himself, his own staff and with his own equipment. Some items cannot be costed on unit basis and it is then normal to estimate the time to be taken on a worker day basis together with the use of any tractor and equipment needed, eg to remove the concrete base of a former cow kennel building (a tenant's fixture) sold off by the previous tenant, but leaving the base, would take, say three worker days for a farmer, together with three days' use of a tractor, loader, bucket and trailer. The claim would then be calculated as follows:

3 worker days @ £88.55	£265.65
3 days tractor @ £77/--	£231.00
2 days FE loader and earth bucket @ £30	£60.00
1 day trailer @ say @ £25	£25.00
	£581.65

11.7 Hedges and fences

Hedges become frequently overgrown, high, wide, gappy and not stockproof. Often they are in need of cutting and laying. Some hedges, eg hedges with a predominance of hazel and young thorn, are much easier to cut and lay than others.

Others are very difficult to lay, with many strong years' growth that needs considerable axe work to lay and also cutting out with chainsaws. Note that where the tenancy agreement specifies cutting and laying a proportion of the hedges each year, this should be done. If there is no tenancy agreement, SI 1973 No 1473, paragraph 9 operates, and this requires the cutting, trimming or laying of a proper proportion of the hedges in each year in order to maintain them in a good and sound condition.

Note, that the said paragraph 9 specifies, as do many tenancy agreements, that a proper proportion of hedges should be laid annually. Others are trimmed and some left to grow for laying at a future date.

Occasionally, a tenancy agreement may specify that a tenant shall plant up gaps that have developed, with quickthorn. Where this is not specified and gaps exist, a claim will arise for gapping up with deadwood or possibly railing the gap or probably making the gap stockproof with posts and wire.

Woven wire fences deteriorate after a number of years (usually 15), the wire rusts and becomes loose and posts rot, frequently part or the whole need replacing. In districts suffering from atmospheric pollution, or near the seaside, wire will rarely last more than 10 years.

Post and rail fencing again rots and becomes defective and needs replacement. Barbed wire is usually a tenant's fixture but where this is established to be the landlord's property, a claim will arise if posts rot and the wire becomes loose and rusts.

Hedges should be trimmed annually (where they are not left for growing on, to lay or the farm is an environmental scheme (with landlord's consent)).

Frequently brambles and briars etc encroach onto the adjacent field. A claim can arise for cutting these back.

Specimen claims

High very strong, wide and gappy hedge
Cost of cutting and laying . . . m @ £

Overgrown hedge
Cost of cutting and laying . . . m @ £

Hedge cut down with Shapeshaw but not stockproof
Cost of erecting alongside the hedge line a woven wire fence on tanalised or pressure treated posts at 2m centres with single strand of barbed wire . . . m @ £

Hedge has numerous gaps
Cost of planting two rows (staggered) quickthorn and protected on both sides with woven wire fencing with barbed wire strand . . . m @ £

Pig wire fence rusted, weak and not stockproof
Cost of removing and replacing with post and medium gauge pig netting fence with two strands of barbed wire . . . m @ £

Post, four rail fence defective in places and not stockproof
Cost of removal of defective sections and replacing with sawn tanalised replacement fence to match existing . . . m @ £

Two strand barbed wire fence rusted and defective
Renew . . . m @ £
Single strand renew . . . m @ £

Hedge not trimmed
Trim and dispose of trimmings . . . m @ £

Brambles, briars and other undergrowth encroaching at sides of hedge into adjoining fields
Cost of cutting with flail cutter . . . hrs @

11.8 Gates

Tenants are usually responsible for maintaining and replacing gates. Some old tenancy agreements still provide that the landlord should provide materials for gates, in which case a claim would be restricted to the cost of making and erecting the gates. Some old gateways are still 10ft wide which is too small for modern equipment. The outgoing tenant will not be responsible for providing a larger gate nor will he be responsible for providing better quality gates than previously existed, eg a 15ft (4.55m) galvanised iron gate in place of a 10ft (3.03m) wooden gate. Gates should be complete with all hanging and securing fittings. Where gateways are depressed by stock traffic and a large gap exists under the hung gate, a claim can arise for building this up to its proper level with earth or stone, as required.

Gates are frequently bent, broken, are not properly hung and drag on opening, and sometimes fittings are missing. Again, the hanging and shutting posts may need replacing and re-setting and side rails need repairing or replacing.

Specimen claims

3.66m (12ft) metal gate damaged beyond repair
Remove and replace with similar gate complete with fittings and properly hung.

Gate opening existing but 3.66m (12ft) gate and posts missing
Supply and erect 3.66m metal gate complete with hanging and shutting posts, together with all necessary fittings.

Top rail of gate and shutting post bent
Straighten out top rail and replace shutting post complete with catch.

Gate not hung properly
Take up loose hanging post, reset and rehang gate to close properly.

Securing catch to gate missing
Replace.

11.9 Ditches, drains and culverts

Ditches frequently become filled in, blocked, silted up, weeds grow on the sides and on ditch bottoms. Where not protected by barbed wire at the side, stock tread the sides in. Occasionally drain outfalls discharging into ditches also get trodden in and partly blocked. The remedy is to clean the same out and to protect (if stock is kept on the adjacent land) with a barbed wire fence. It should be noted that barbed wire fencing against ditches cannot be claimed for, unless it existed at the commencement of the tenancy. In times gone by, ditches were cleaned out by hand (and this may be necessary even today, where tractors cannot approach the ditch) but nowadays they are generally cleaned out, mechanically. The work is normally carried out by specialist contractors.

Culverts and bridges become damaged and occasionally need rebuilding and sometimes repairing.

Tenants are usually responsible for maintaining field drains and where the agreement is silent or does not exist, SI 1973 No 1473 provides that he shall be responsible for keeping them clear of obstructions. Blockages can sometimes be located and made good and the cost of making good is the amount of the claim. However, it occasionally will be cheaper and more effective to lay a new drain in a wet area than try to locate a blockage.

Specimen claims

Wide water, course (average width 0.5m) blocked and in need of cleaning
Cost of cleaning out deep wide watercourse and spreading soil . . . m @

Culvert (4m) fallen in
Dig out and renew providing 3 no . . . mm dia. concrete pipes surrounded in consolidated shingle and finished with 6in deep concrete surface

Ex-railway sleeper bridge (4m) rotted and weak
Renew — 2 w/d @
    14 sleepers

Field drain blocked
Cost of unblocking — 1 w/d

Wet area where drains not functioning
Cost of laying 80mm plastic pipe drains 50m @

11.10 Foul land

Weed infestation can take several forms such as:

(a) couch (or squitch) infestation in arable and pasture land
(b) black grass in cereals
(c) docks in pasture and arable land
(d) wild oats in cereals
(e) bracken in pastures
(f) annual broadleaved weeds eg thistles, mayweed, poppy, charlock, etc in arable land.

(a) Couch

Most frequently this is found in cereal crops, although it infests any arable crops and is also found in pastures. The expensive additional cultivations and fallows of yesteryear are no longer necessary. Treatment is by spraying stubble with Roundup at 3 litres per ha. This costs approximately £15 per ha, this being done when the weed is growing profusely after harvest. It is not necessary to leave the spraying until after harvest as pre-harvest spraying, at 3 litres per ha depending on the density of couch and can be done provided the crop is at the correct stage of maturity.

The use of additives to reduce the cost of Roundup application is commonplace today. Roundup Gold contains its own additive and increases chemical uptake of the couch (£5 per litre). If pastures are infested, complete kill of all couch and other grasses will be effected by application of Roundup at 4.5 litres per ha costing £22.50 per ha.

(b) Blackgrass

Blackgrass infestation is an ever increasing problem in the main cereal growing area of the UK. Its control within cereal crops is becoming more difficult as herbicide resistance becomes commonplace within areas of high blackgrass populations where perhaps continuous cereals have been grown.

Control of blackgrass now starts with good cultural control (stale seedbeds), regular break crops to enable different chemistry to be used (Kerb in OSR for example) and a robust pre-emergence herbicide. Control programmes in winter wheat can cost up to £65.00 per ha for moderate to high populations of blackgrass.

(c) Docks, thistles and nettles in pasture

All three are found in both permanent and rotational grass. There are specific chemicals for each however the use of Pastor or Forefront will control all three at around £40.00 per ha. For general broad leaved weeds MCPA/CMPP can be used costing £15.00 per ha. The presence of clover within grass swards should be determined as all the above treatments will kill or severely check clover.

(d) Wild oats

Wild oats have spread the breadth of the country, probably through purchased seed grain. To make a claim for wild oat infestation it will be necessary to inspect the crops on the farm concerned during the months of June to August, prior to the termination of the tenancy, when the infestation is apparent. However, care should be taken to find out whether the wild oats have been rogued, ie picked out by hand, prior to inspection. Wild oat seeds can remain dormant in the soil for years. Distinction should be made between wild oats and the occasional cultivated oat plants (from a previous crop) which are sometimes seen growing in a different crop and which may be odd seeds that have germinated. Wild oat plants have awns (or spikes) at the end of the seed whereas oats do not.

Wild oats can be controlled in all crops using different chemicals. Triumph/Cheetah in Winter Wheat will cost £17 per ha, Axial in Winter Barley will cost £20 per ha. Where bad infestations are expected a pre-emergence herbicide can be used based on Pendimethalin on Tri-allate based products costing approximately £32 per ha.

(e) Bracken

Where this is a problem, as is occasionally experienced in upland and rough pastures, it can be controlled by spraying with Asulox at a rate of 11 litres per ha @ £8 per litre. It can be aerially sprayed but the cost will be much greater.

(f) Annual broadleaved weeds

There are a host of broadleaved weeds that affect cereals. In most parts of the country control is found by using Pendimethalin, Iosproturon or Diflufenican mixes during the autumn. In the advent of a ban on chemicals containing Isoproturon the emphasis will be on Pendimethalin (Stomp) mixes and any new chemistry released on to the market. General costs will be £21–£25 per ha.

The application of an Ally/Harmony M type product can be used in the spring at a cost of £25 per ha.

Specimen claims

Thistle and general weed infestation in pasture
Spray once with MCPA or MCPB @ . . . ha

Couch infestation after winter wheat crop
Cost of spraying with Roundup @ £ . . . per ha

Nettle infestation in pasture
Cost of spraying with

Dock infestation in pasture
Cost of spraying with

Annual weed infestation including thistle, charlock and mayweed in winter barley crop
Cost of spraying with

Severe wild oat infestation in all crops
Cost of two annual successive sprays with

11.11 Pasture repair

11.11.1 Often cattle are outwintered and cause damage to pasture and, in particular, poaching occurs at gateways, feeding areas and also sometimes over the whole field through hoof marks, tractor and trailer wheel marks to cattle racks, etc. Often the damage resultant is not sufficient to warrant ploughing up and re-seeding. The damage can usually be rectified by eg two disc harrows to the affected area, applying fertiliser, broadcasting grass seed, a light harrow and a rolling. Nevertheless, a loss will also result to the incomer in that this will take time to establish and he can suffer loss of grazing in the meantime. An example of the work necessary and its cost (2008) is:

	per acre	per ha
2 disc harrows	17.00	42.00
1 fertiliser spinner	3.15	7.80
1 seed broadcast	3.15	7.80
1 light harrow	6.30	15.50
1 roll	4.10	10.10
Seed	35.00	86.48
Fertiliser 2 × 50kg 20:10:10 @ £160	16.00	39.53
Electric fencing area off say	5.00	12.35
	£89.70	£221.56

But since this area is a small one, costs should be increased by 30% to say	£116.61	£288.02

If the damage is severe, it may be necessary to sub-soil the affected area.

Occasionally small rutted areas are found and these, as a rule, can be repaired by disc harrowing or rotovating, broadcasting seeds and rolling.

Specimen claim

2ha pasture poached and to be re-established
Cost of cultivations, seeds, fertilisers and fencing off
2 ha @ £
Loss of grazing during re-establishment
4 weeks spring grazing @ £ . . . per ha/per week

11.12 Dilapidations to buildings etc

11.12.1 A thorough knowledge of the Law of Dilapidations is necessary. As far as agricultural tenancies are concerned, the case of *Evans* v *Jones* [1955] 2 QB 58 is relevant where the tenancy has only existed for a short time. Here it was held that the tenant could not have reasonably been expected, during the short term of the tenancy, to have to put the premises in repair, having regard to the condition at the outset and this must be taken into account when assessing the landlord's entitlement to dilapidations where the tenancy is of a short duration. If the tenancy has existed for a long period, this rule will be of no help to a tenant.

11.12.2 In assessing building dilapidations, it is essential that all items should be measured up accurately and priced out in the proper way using an up-to-date builders' price book.

11.12.3 The claim will be assessed under the terms of the tenancy agreement or under SI 1973 No 1473 dependent on whether the claim is under section 71(1) or section 71(3) of the 1986 Act. If under SI 1973 No 1473, its provisions must be studied very carefully and in particular paragraph 11, which clearly states that any accrued liability on the part of the tenant in respect of internal and external decorations can be the subject of claim.

11.12.4 Assuming repairing liabilities are in accordance with SI 1973 No 1473 the tenant is responsible for a large number of repairs including, *inter alia*, boilers, fireplaces, drains, manholes, water supply systems (above ground), tanks, troughs, pumping equipment, cattle grids, bridges, ponds, roads and yards. He is also to keep clean and in working order roof valleys, gutters and rwdp wells, septic tanks and cesspools and also to pay half the cost of repairs to doors, windows, interior staircases, eaves guttering, downpipes, floorboards and skylights. Note, however, that by virtue of 1(3) Part I of the SI, the tenant is not responsible for replacing items mentioned in 5(1) Part II, eg boilers, ranges, water supply systems, tanks, pipes, etc, which have worn out or become incapable of further repair, unless the tenant is responsible for its replacement under paragraph 6 (where the replacement is necessary due to the wilful neglect of the tenant etc). Reference is also made to the case of *Robertson Aikerman* v *George* (1953) 103 LJ 496 in which it was decided by Leicester County Court that the landlord could not recover from the tenant the one half cost of repairs of floorboards, doors, etc, unless the work had been actually done at the termination of the tenancy. It appears, therefore, that the landlord should ensure that work, where he would normally recover half cost, should be undertaken prior to the termination of the tenancy.

Specimen claims

Central heating boiler not functioning
Repair and put in working order

Ball valve to automatic water trough in yard leaking
Repair and put in working order

5 tubular bars to cattle grid at farm entrance bent
Take out and renew

Right hand stone parapet to bridge on farm drive damaged
Take down and re-build

All eaves and gutters on south elevation of farmhouse blocked
Clean out gutters, and leave in working order

Septic tank to farmhouse overflowing
Clean out and leave in working order

Farm drive severely potholed
Cut out potholes and fill with tarmacadam, properly consolidated to
level of surrounding road surface

Lounge in need of redecoration. Redecorated over 7 years ago
Redecorate including all preparation work, stripping wallpaper, stopping
and making good, rubbing down, preparing and applying 2 coats oils
to all previously painted wood and iron work, twice whitening ceilings
and re-papering walls — say, if redecorated say 4 years ago the claim
would be £ . . . × 4/7 = £

No 2 windowpanes cracked in bedroom No 1
Hack out, prepare rebates and reglaze in 4mm sheet glass — 0.4 m2 @ £

*No 11/3 m2 of boarded floor to Granary rotted due to tenant keeping poultry in
the building*
Take up defective area and renew m @

11.13 Other claims

Other claims may arise contractually, and under section 11 of the
Agriculture Act 1947. These may embrace the following claims.

11.13.1 Shortage of new seeds

This claim may arise both under section 11 and also contractual
obligation. If a landlord's interest is damaged through such seeds not
being left, the claim will be for making good such shortage. However, in

computing such claim, consideration should be taken of the possibility of land growing such seeds being available for growing other crops.

11.13.2 Departure from rotation

The tenant has generally a freedom to crop arable land as he wishes (section 15 of the 1986 Act), except in the last year of the tenancy. However, if this system of cropping causes injury or deterioration to the holding, the landlord can recover damages under section 11 of the 1947 Act and also both sections 71 and 72 of the 1986 Act.

11.13.3 Loss of quota

No claim can arise here except where there is a specific breach of the contract of tenancy or loss of quotas which the tenant cannot dispose of without his landlord's consent.

11.13.4 Selling produce in the last year of tenancy

The tenant is not allowed, by virtue of section 11 of the 1947 Act, to sell off or remove produce in the last year of tenancy. If so removed, the landlord can claim for the equivalent manurial value of the crops sold off. This applies only to produce normally consumed on the holding such as fodder but not corn.

11.13.5 Continuous cereal growing

Damage can be caused to the freehold by continuous cereal growing resulting in infestation of wild oats, couch, blackgrass, pests, loss of fertility, disease, etc. The soil structure may also be damaged. Claims may arise under both section 71 and 72 of the 1986 Act.

11.13.6 General

The landlord's claim in respect of the above and possibly other matters may not be properly compensated under section 71, in which case claims should also be made under section 72 for deterioration. However, for such a claim to succeed it is reminded here that a proper notice of intention of claiming under this section should be served before the expiration of one month before the tenancy terminates. The compensation recoverable for a deterioration claim is restricted by virtue of section 72(3) to the decrease in the value of the holding, as a holding having regard to the character and situation of the holding and the average requirements of tenants skilled in husbandry.

Example of Dilapidations Claim

AGRICULTURAL HOLDINGS ACT 1986
Re: The Holding known as:
WEST FARM, METCHINGFIELD, HEREFORD

To: Mr J. Yeoman-Farmer
       Holly View
       Dinedor
       Hereford
       (Outgoing Tenant)

WE HEREBY GIVE YOU NOTICE pursuant to Section 83(2) of the above Act of our intention to make against you certain claims arising out of the termination of the tenancy of the above holding the nature of which claims is set out in the Schedule hereto.

THE SCHEDULE

DILAPIDATIONS:
Claims made under clauses 111(4)(9) (10)(19)(27) of the Tenancy Agreement dated 4th April, 1972 and also under SI 1973 No 1473

FARMHOUSE

1.	All eaves gutters require cleaning out. Clean out.	£25
2.	Septic tank full. Clean out.	£75
3.	Bedrooms No 5 (NE) and 6 (SE) require redecoration. To cost of stripping walls, making good, preparing and painting all wood and iron work previously painted two coats oil and re-papering walls with comparable quality paper.	£1,200
4.	WC pan cracked in Bathroom. Replace.	£60
5.	Hot water tap in Cloakroom leaking. Replace washer.	£10
6.	Garden. Vegetable garden very unkept and weed infested. Dig, leave tidy and control weeds.	£120

FARM BUILDINGS

7.	Dairy No 5 window panes broken. Renew.	£50

8. Range of 3 Calving Boxes
- (a) No 2 water bowls missing. Replace and re-connect supply. £75
- (b) Half heckdoor to middle Box damaged beyond repair. Cost of replacement inc. fittings £100
- (c) Gulley grid missing. Replace £25
- (d) Manhole cover cracked. Replace £60

9. Range of No 4 Calf Cots
Limewash to walls to renew. Cost of cleaning and carrying out necessary work. 2 W/D and Materials £200

10. Barn
Renew missing and loose slates Cost of renewal (Maximum permitted) £100

11. Pump House
'Hydra' water pump not working. Cost of putting into working order. (Estimate attached) £180.26

12. Slurry Lagoon
Full up. Cost of cleaning out. 4 W/D (with agitator, tractor, pump and slurry tank and spreading) @ £150 £600

13. Electrical
No 1 switch in Piggery broken and No 3 electric pendants in covered yard missing. Cost of putting into repair. £48.20

14. Farm Drive
Severely potholed in many places. Cost of cutting out and filling with tarmac and rolling £625

LAND

Pt OS No 602 8.79ha Winter Barley

15.	602/SE	High overgrown hedge. Cost of cutting and laying 323 m @ £8.50	£2,745.50
16.	602/SE	Wide ditch blocked. Cost of cleaning out 310 m @ £2.00	£620.00
17.	602/652	Culvert broken. Cost of reconstructing.	£200.00

Pt OS No 652 3.32ha Winter Barley

18.		Severe infestation of wild oats. Cost of spraying with Avenge in 2008 and 2009. 3.25 ha @ £50	162.50
19.		Land drain blocked in SE Corner. Cost of locating blockage and making functional	£50
20.	652/220	108m run of gaps in hedge. Cost of making stock proof @ £4	£432

21.	652/220	Top bar of metal gate bent. Cost of making good.	£15

OS No 220 7.42ha Winter Wheat

22.		Severe infestation of couch. Cost of spraying with Round up after harvest of 2008 and cost of ditto spraying with half strength dose in 2009. 7.25ha £30	£217.50
23.	220/218	Post and rail fence rotted. Cost of removal and renewal 156m @ £11	£1,716

OS No 218 6.87ha Pasture

24.		Infestation of docks to approx 2ha in NW corner. Cost of spraying twice @ £36	£72
25.	218/618	Hedge not trimmed and not left for growing to lay. Cost of trimming 2 hrs @ £18	£36
26.	218/216	No 2 Hedgerow Oak Trees with barbed wire nailed thereto. Loss of timber.	£50

OS No 618 4.56ha Roots and Stubble

27.		Uncultivated area of 0.5ha in SW corner (site of dung heap) full of weeds. Cost of eradicating weeds, cultivating and bringing back into proper production	£60
28.	618/653	Dry stone wall fallen into disrepair. Cost of rebuilding 10m^2 @ £40	£400
29.	618/653	Gate opening where gate previously existed but no gate or posts. Cost of replacing 3.6m metal gate and posts all complete	£170

OS No 832 5.25ha Ploughed

30.	832/618	Overgrown hedge. Cost of cutting and laying 289m @ £8	£2,312
31.	822/618	Area of briars encroaching into field at headland at S end. Cost of cutting back and burning 1 W/D	£100
32.	832/618	Shutting post missing to gate. Cost of replacing	£30
33.		Pond in SE corner silted up and outlet drain blocked. 8.5 hours with mechanical digger and 2 hours tractor and trailer	£370

OS No 54 9.10ha Pasture

34.		Whole field infested with thistles. Cost of spraying @ £90 per ha	£819
35.	54/53	Metal gate damaged beyond repair by vehicle collision. Cost of replacing and hanging	£95
36.	54/53	Side rails both sides gate opening disrepair. Cost of removing and replacing 6m @ £10	£60

37.	54/53	Small ditch silted up. Cost of cleaning out and spreading spoil 171m @ £1.50	£265.50
38.	54/53	Landlord's barbed wire fence alongside ditch in disrepair. Cost of re-erecting and replacing No 25 fencing stakes	£60

OS No 651 7.03ha Pasture

39.		Approximately 4 ha of Pasture infested with clumps of nettles. Cost of spraying and restoring infested areas to proper production	£240
40.		Whole field infested with thistles. Cost of spraying £18 per ha	£126
41.	651/654	Lane on boundary to part overgrown. Cost of clearing and controlling weeds	£120

OS No 654 5.68ha Pasture

42.		Approximately half infested with docks. Cost of spraying with Pastor two times to eradicate and loss of production	£200
43.		Approximately 1.75 infested with couch on N boundary alongside hedge. Cost of spraying	£90
44.		Approximately 1ha severely poached by outwintered cattle. Cost of restoration and loss of production	£200

RESERVATION

The right is reserved to amend the above claim at any time prior to settlement or arbitration.

Dated this fifth day of February 2008

. . . .
WILLIAMS, DAVIES & REES, Agricultural Valuers
DOLANOG
WELSHPOOL
POWYS
Duly Authorised Agents and Valuers on behalf of the Walford Estate Company.
Landlords.
E&OE

PRACTICE NOTE:

ITEM 25: Since this is the only hedge to trim and having regard to the extra costs of getting a contractor onto the farm to do this work, the normal charge is doubled.

Pipelines, Capital, Easement Payments and Claims

12.1 General

Claims arise following compulsory rights orders to lay sewers and mains water under the Water Industry Act 1991, gas mains (under the Gas Act 1986 — section 70 and schedule 3 as amended by the 1995 Gas Act), commercial pipelines (eg oil) under the Pipelines Act 1962. Three types of claim arise:

(a) easement claims for gas and commercial pipelines
(b) capital payments for sewer and water mains (recognition payments)
(c) damage claims arising following the laying of any pipelines.

12.2 Capital payments

These arise following the laying of sewer and water mains and are a recognition payment for the presence of the pipeline as no easement is taken. In the past it has been the practice of the authorities to pay 50% of the capital value of the freehold strip affected. Water authorities usually contended that this should be restricted to narrow notional easement widths occupied by the pipes usually from 3.5m to 5m. However, the case of *St Johns College Oxford* v *Thames Water Authority* [1990] 1 EGLR 229 decided that payment was to cover 50% of the full working width of 20 yards and held that the damage on injurious affection caused by the pipe affects not mainly the strip but each farm as a whole.

Additional payments are made for marker points (usually £10 each) and also for chambers, valves, manholes and other permanent structures. If these manholes etc are in arable fields they are a serious nuisance and it is not unusual especially in the case of the larger chambers for compensation from £100 to £500 each to be paid for their presence. If the chambers are placed in hedgerows the interference is less, as is the compensation eg £50 to £70 each. Sometimes an authority can be persuaded to bury a manhole below the depth of cultivation and this will result in lesser compensation payable.

12.3 Easement payments

These payments arise when commercial pipelines are laid, eg oil and for gas mains. National Grid and promoters of commercial pipelines normally pay an easement payment of 75% of the vacant possession value of the strip of land affected. Quite often, dependent on the width of easement these have been up to £30 per m run. The higher payments are usually in respect of small parcels of land or paddocks attached to dwellings. Larger farms have been compensated in 2007 generally where the land is of top quality at the capital value of £6,000 per acre and in some cases more.

In addition an occupiers payment is made to the occupier (whether the owner or tenant) hopefully to obtain co-operation. This payment is normally agreed between the promoters and the CLA & NFU.

Again payments are made for manholes, chambers and other structures.

12.4 Damage claims

These should cover the following items.

(i)   Actual crop losses.
(ii)   Future crop losses — over a number of years from, say two years to 10 years, dependent on damage caused when the work was done.
(iii) Cost of restoration.
(iv)   Cost of replacing lost fertility.
(v)   Weed control. Weeds have a habit of spreading on pipe-tracks and also being introduced to clean land where they did not previously exist.

(vi)   Cost of future hedge-laying costs where hedges are re-planted. This is usually in 10–12 years' time.

(vii)   Cost of replacing top soil where subsidence occurs after a period of time. Also for replacing any top soil lost.

(viii) Drainage. Frequently drains are cut. It is important that they should be properly reinstated after the main is laid. In small diameter pipelines, a generally satisfactory method is for a section of plastic drain pipe to be laid on a substantial 100mm × 100mm hardwood baulk which is let into the soil for 0.5m either side of the trench, and connected to the exposed drains. In the case of large diameter pipes, some support such as concrete built up from the main to support the actual replacement drain, is needed. In any event, the authority laying the main should be held responsible at all times for any damage caused to drains as a result of their operations.

(ix)   The claimant should be paid for all the time he spends on the matter, eg with agents, contractors, valuers, solicitors, rounding up stray stock, telephone calls, travelling expenses, etc. A detailed diary of time used should be kept.

(x)   All professional fees.

(xi)   Interest on the agreed claim. However, it is usually difficult to recover this from the pipe-laying authority, except on the easement payment.

Example easement recognition payment

A 283m run of 150mm diameter water mains is laid through a farm. The working width is 15m. There are four marker points and three valve chambers (one in a hedgerow). The suggested claim is:

(a) Pipeline
283m × 15m – 4245m² or 0.424 ha
0.424h @ £14,825 (£6,000 per acre) × 50%   £3,143

(b) Marker Points
4 No @ £10   £40

(c) Valve Chambers
3 No @ £400   £1200
1 No @ £60   £60   £1,260

(d) **Fees**

Valuers' & Solicitors' Fees

(e) **Interest**

On the claim from date of entry to date of payments.

Example of Pipeline Claim

STATEMENT OF CLAIM
by
MR RJ BENGOUCH

against
THE VALE OIL CORPORATION LIMITED

Following the laying of an Oil Pipeline through
WEST FARM, METCHINGFIELD, NORFOLK

March 1998 Prepared by: Williams Davies & Rees
Agricultural Valuers
Dolanog
Welshpool, Powys

1. LOSS OF CROP
1.1 OS 1481 — Working Area 0.24ha

(a)	2007 — Loss of Copain winter wheat 9 tonnes per ha @ £180 per tonne (on seed contract)	388.80
(b)	2007 — Loss of straw 3 tonnes per ha @ £35 per tonne	25.20

1.2 OS 8352 — Working Area 0.30ha

(a)	2007 — Loss of Desiree potato crop, 50 tonnes per ha @ £150 per tonne	2,250
(b)	2006 — Loss of barley straw 2.5 tonnes per ha @ £40 per tonne	30

1.3 OS 1545 — Working Area 0.4ha

(a)	2006 — Loss of sugar beet crop Site works made it impossible to harvest the sugar beet in the area of the field south of the easement. Total area lost 1.58 ha 54 tonnes per ha @ £21 per tonne (net of transport)	1,791.72
(b)	Loss of sugar beet tops 1.58ha @ £75 per ha	118.50

(c)   2007 — Loss of Triumph spring barley
5 tonnes per ha @ £150 per tonne   300

1.4 OS 2400 — Working Area 0.42ha

(a)   2006 — Loss of winter grazing @ £99 per ha   41.58
(b)   2007 — Loss of spring, summer and autumn
grazing @ £346 per ha   145.32

1.5 OS 6633

2006 — Loss of Avalon winter wheat. Site works made
it impossible to harvest area of crop to the south of the
easement. Total area lost 0.121ha. Loss of 10 tonnes
per ha @ £161 per tonne (on seed contract)   194.81
Loss of straw   15.00

1.6 OS 9500

2006 — Works to the Main caused a loss of 0.2ha of
sugar beet in the corner adjoining OS 1545 and OS 8351
Loss of 54 tonnes per ha @ £21 per tonne   226.80
Loss of beet tops @ £75 per ha   15.00

1.7 Loss of Area Single Farm Payment

(a)   2006 — 0.68ha @ £251   170.68
(b)   2007 — 0.68ha @ £240   163.20

2.   REINSTATEMENT OF HEDGES

OS 8352/9500	18m
OS 9500/1545	15m
OS 1545/2400	30m
OS 2400/6633	18m
OS 6633/x	18m
OS 4500/Railway	16m
	115m

To cost of first laying of re-planted hedges in 10 years.
115m @ £13 (allowing for inflation and deferring)   1,495

3.   REINSTATEMENT OF WORKING AREA
3.1 OS 4500, 8352, 1545 and OS 633 — Total area 1.01ha

To cost of additional expenses in cultivating and working
down the area affected by the laying of the main 1.01ha
@ £100 per ha   101

3.2 OS 2400 — Area 0.42ha

To cost of reinstating the working area including the levelling
of the surface, panbusting, cultivating, and re-seeding
(two sections) 0.42ha @ £494 per ha   207.48

3.3 Restoration of Fertility

The Corporation to pay for the cost of restoring fertility
to the areas affected by the laying of its main. This involves
the application of compound fertiliser, lime and farmyard
manure. The claimant has had the trench area tested
for acidity and a pH value of 4.8 has resulted. It is
recommended that 12.33 tonnes of lime per ha be
applied in two doses over two years.

12.33 tonnes lime @ £15 per tonne	184.95	
25 tonnes FYM @ £15 per tonne	375.00	
12 × 50 kg 21:8:11 Compound @ £260	156.00	
1.43ha @	715.95	1,023.80

3.4 Removal of Surface Debris

The Corporation to pay for the cost of removal of rock
and other debris brought to the surface following the
laying of the main.
3 worker days plus tractor and trailer @ £177.30 per day   531.90

3.5 Removal of Wire Fencing

The Corporation to pay for the cost of removing the
wire fencing marking the boundary of the working
area 2 worker days plus tractor and trailer
@ £177.30 per day   354.60

3.6 Weed Control

The Corporation to pay for spraying the affected area with
a herbicide to control the inevitable growth of weeds
1.43ha @ £85 per ha   121.55

4. REINSTATEMENT OF DISUSED RAILWAY LINE

4.1 Repairs

The old railway line has been damaged by the contractors
and is now very rutted and holding surface water. The bank
against OS No 4500 is badly reinstated and a large heap
of earth left. To cost of reinstatement including levelling
and stoning   500

4.2 Fence

A double set of post and rail fencing should be provided between the old railway line and OS 4500 and planted with a quick thorn hedge, or alternatively the Corporation to pay compensation 15m @ £50 per m   750

5.   CONSEQUENTIAL LOSSES
5.1 Sheep Damage

(a)   Sheep got out of OS 2400 and OS 1545 on three occasions due to the inadequacy of the fence to the working area. They entered OS 9500 and ate the emerging sugar beet plants. Losses on 4 ha of sugar beet estimated at 30 tonnes @ £21   630

Loss of sugar beet tops estimated   100

(b)   The sheep also disturbed the pre-emergent spray in OS 9500 and it did not work, resulting in a weed problem To cost of 2 steerage hoes on 8 ha @ £29.30 per ha   234.40

(c)   The sheep grazed the pasture in OS 6235 in the winter, but due to the inadequacy of the fencing to the working area they escaped into OS 4500 where they grazed some 8 acres of Avalon winter wheat.
Loss of 2 tonnes @ £180 per tonne   360

(d)   One 3-year-old mule ewe fell into the trench and died

To loss of ewe   75

Loss of lambs, 180% lambing percentage @ £60 per lamb (early lambing)   108   183

5.2 Damage to Roadway

The Contractors used the farm track in the winter and damaged the same.

To cost of making good track including labour   600

To charge for allowing contractors the use of the track   300   900

5.3 Disturbance to Rotation

(a)   The farm rotation has been disturbed in OS 8352, 8358/7832, making a total of 6ha. Due to this, Triumph spring barley had to be planted instead of winter wheat. A loss occurred on this disturbance to the rotation. Spring barley yielded 5 tonnes per ha as against a wheat yield of 9 tonnes per ha
Loss of 4 tonnes per ha @ £180 per tonne   4,320

(b) The trench excavation prevented flood water draining away in OS No 4500 — Copain winter wheat. As a result, 1.6ha were lost and had to be re-drilled with spring barley.
Loss of 6 tonnes @ £180 per tonne plus cost of resowing   1,080

6. DISTURBANCE LOSSES
6.1 OS 1545 — Sugar Beet
Due to laying of main it was additionally expensive to harvest (owing to extra time turning of harvester, etc) the sugar beet in OS 1545 To additional cost   250

6.2 Disturbance
The Corporation to recompense the claimants for disturbance and inconvenience suffered resulting from the works. The sum of   800

7. LOSS OF FUTURE YIELDS AND PROFITS
Due to the very difficult working conditions arising from the wet autumn of 2006 and the operations generally in 2007 serious damage was caused to the soil structure with the result that the restoration works proposed will not fully indemnify the claimant from future losses of yield and consequently profits:

1st year 60% of £750 per ha	450	
2nd year 30% of £750 per ha	225	
3rd year 10% of £750 per ha	75	
1.43 ha @	750	1,072.50

8. EASEMENT PAYMENT
The Corporation to pay the claimant compensation as follows:

8.1 Pipeline Easement (Freeholder)
931 m ×20 m = 1.862ha
1.862ha @ £14,826 (£6,000 per ha) × 75% =   20,704.50

8.2 Marker Points
6 no @ £10   60

8.3 Chambers

No 3 Chambers @ £200	600	
No 1 Chamber @ £50	50	650

8.4 Occupiers' Payment
At the scale agreed with the NFU

9. ADDITIONAL ITEMS
9.1 Drainage
The Corporation to make good at all time damage caused by the operations to the land drainage system, whether artificial or natural, and also to pay for losses caused by such damage and also by the restoration work put in hand.

9.2 Subsidence
The Corporation to make good any subsidence resulting from the mains laying operation, including provision of a minimum of 0.3m depth of good quality top soil of a quality equivalent to that lost where necessary.

9.3 Fees
The Corporation to pay the claimant's valuer's fees and expenses on the approved scale, together with solicitor's costs.

9.4 Interest
The Corporation to pay interest at the statutory rate on the agreed claim from the date of entry.

9.5 Any Other Matters

E&OE.

Practice Note
As a rule the easement payment and payment for chambers and the occupier's payment is agreed before entry on to the land. Marker point payments are usually dealt with at the time. However, they are shown here in this specimen claim to illustrate the type of claim made.

Electricity Line Wayleaves and Claims

13.1 General

Substantial compensation claims usually arise on the construction of 400 kv and 132 kv overhead transmission lines. The former are usually erected by the National Grid and the latter erected or maintained by area electricity boards, who are also responsible for smaller transmission lines. Claims take two forms.

13.2 Losses arising as a result of constructional work

This claim covers crop losses, future crop losses and all other losses arising on the line being erected. This also covers timber cut down, trees lopped, and damage to soil structure (frequently this arises as a result of heavy equipment severely rutting and consolidating the area concerned). A claim will broadly follow the specimen claim given in respect of the laying of a main.

13.3 Damage to the freehold

Electricity lines are unsightly, mar the landscape, interfere with views from farmhouses and cottages, interfere with irrigation systems and sporting rights, and generally depreciate the value of a freehold. These items cannot, as a rule, be claimed for unless the claimant enters into a deed of grant, granting a perpetual easement for the line in return for compensation. The claim is usually assessed as follows.

(a) The annual rental payment is capitalised by multiplying × 20 YP. (Where there is a second line in the same field of 20 acres or less, the capitalisation is increased by 50%.) This is usually the arable rate of payment unless it is clear that the land concerned is not arable land and cannot be ploughed, eg hill land.

(b) Dwellings are valued and if the line is near and the view marred, a percentage is agreed as being the depreciation in value, eg a farmhouse may be worth £500,000 and the depreciation is agreed at say 5% or 10% or more. Cottages may be assessed in a similar way. The practical difficulty is generally in agreeing the percentage depreciation.

(c) Sporting rights may be seriously interfered with, eg a line going through an excellent shoot or over some good fishing pools on a river. This should be assessed as a reduction in rental, which reduction is capitalised at 25–30 YP.

(d) Irrigation cannot be practised under overhead transmission lines for obvious reasons. Not every farmer irrigates nor has the need to, but if he does the interference and loss can be considerable, as irrigation is normally only used for high value cash crops such as potatoes and market garden crops. An estimate of the average annual net loss should be made, over the area affected, and this capitalised by such a YP as would be used for valuing a farm, say, 25 to 30. However, this is a most contentious and arguable claim.

(e) Where a line passes through a wood, all timber is cut and the area affected is perpetually sterilised, ie trees cannot be grown on the strip concerned, which can be as much as 20m wide. Total losses for the future, arising from this should be claimed, ie the profit on a crop of trees and also the danger to the surrounding woodland from windblow as the canopy has been disturbed.

(f) Any other matters affecting the freehold, eg buildings cannot be erected under overhead lines, loss of development values etc.

13.4 Reference is made to the case *George* v *South Western Electricity Board* [1982] 2 EGLR 214 when the Lands Tribunal decided that a claim could be on the elemental approach (as above) or on the overall view, ie taking a percentage depreciation of the agreed value of the holding, and in this case awarded 5% of the value of the holding of £176,600. The two methods should be used to check each other.

13.5 The decision in the 1983 Land Tribunal case of *Clouds Estate Trustees* v *Southern Electricity Board* [1983] 2 EGLR 186 should also be considered. It emerged here that the annual payments or rents agreed between the electricity authorities and the NFU and CLA included nothing in respect of loss of visual amenity, nuisance value, damage to shooting, etc and these, and possibly other relevant matters, could be the subject of an additional claim. Care should therefore be taken that these matters are not overlooked when negotiating a permanent easement claim.

The case of *Naylor* v *Southern Electricity Board* [1996] 1 EGLR 195 awarded compensation of £4,000 for depreciation in the value of a house due to oversailing power lines.

13.6 In every case, solicitor's and valuer's fees, interest at the statutory rate should be claimed, and also, in the case of the damage claim, payment for the claimant's time (which is usually considerable) spent in dealing with the matter.

Notices to Remedy Breaches of Tenancy Agreement

14.1 The law on this subject is contained in the Agricultural Holdings (Arbitration on Notices) Order 1987 (SI 1987 No 710) and schedule 3, Part I case D of the Agricultural Holdings Act 1986.

14.2 The service of an appropriate notice to remedy breaches is a method whereby a landlord who considers his tenant is in breach of the terms of his tenancy can possibly get the breach remedied. However, service of notices of this kind often gives rise to much contention and costs, and should not be embarked upon unless there are good grounds of serving such notice. A very careful study of the law is required to avoid pitfalls, many of which exist. Photographs of some of the defects taken at the time of preparation of the notice can, in subsequent events, be of great help.

14.3 The notice served must be in the prescribed forms. There are two officially prescribed forms to use under the Agricultural Holdings (Forms of Notice to Pay Rent or to Remedy) Regulations 1987 (SI 1987 No 711), namely:

(a) Form 2 — a notice to be served where the tenant is in breach of agreement and is required to do work of repair, maintenance and replacement

(b) Form 3 — a notice where a tenant is required to remedy a breach of agreement and is not a notice requiring work of repair, maintenance or replacement, eg to reside in the farmhouse.

14.4 In serving any notice, the prescribed notes must be attached to the notice. See notes under the example notice (2) and also the different notice (and note beneath) in the case of Form 3.

Where the notice is to do work, a period of less than six months shall not be treated as a reasonable period.

14.5 In a notice to do work, it is important to clearly state, separately, the breaches and also, following on, separately again, the remedy required. The remedy required should be fair, reasonable and practical, as otherwise an arbitrator can be expected, on a reference, to adopt a fair and practical attitude in his award. It would be unreasonable, for instance, to expect a tenant to lay hedges in the summer months. It is always wise to attach a plan to any notice to do work served, and this should be numbered to correspond with the numbers given on the notice.

It is also prudent to state that it is expected that the necessary work shall be executed to a proper standard and in a workmanlike manner.

Example

Landlord's Notice to remedy breach by doing work of repair, maintenance or replacement

AGRICULTURAL HOLDINGS ACT 1986

Schedule 3, Part 1, Case D
Notice to tenant to remedy breach of tenancy agreement by doing work of repair, maintenance or replacement.

Re: The holding known as
PART OF LOWER BARN FARM
GREAT ONEN
MONMOUTH
OS NOS 231, 231 & 228

To: MR ALBERT CHARLES
GREAT ONEN FARM
MONMOUTH

1.   I hereby give notice that I require to remedy within twelve months, from the date of service of this Notice the breaches set out below, of the terms or conditions of your tenancy, being breaches which are capable of being remedied of terms or conditions which are not inconsistent with the fulfilment of your responsibilities to farm the holding in accordance with the rules of good husbandry.

2.   This Notice requires the doing of the work of repair, maintenance or replacement specified below.

PARTICULARS OF BREACHES OF TERMS OR CONDITIONS OF TENANCY

Terms or Conditions of Tenancy Clauses 5(7), 5(8), 5(9), 5(14) etc of the Tenancy Agreement dated 31st December, 1952. Also the rules of Good Husbandry as set out in section 11 of the Agriculture Act 1947 and Statutory Instrument No 1473, The Agricultural (Maintenance, Repair and Insurance of Fixed Equipment) Regulations 1973.

Particulars of Breaches and work required to Remedy them

No on attached plan	OS No of field	
1	231	Pasture Foul — Eradicate couch, agrostis, docks, thistles, chickweed, etc, to bring field back into full production
2	231/243	Nettles, bracken and briars encroaching into field. Eradicate and control to bring area affected back into proper production
3	231/243	Weak gappy hedge. Cut and lay and render stockproof
4	231/199	Ditch overgrown and not cleaned out. Remove overgrowth, shrubs and briars and clean out to proper fall
5	231/199	Weak gappy overgrown hedge. Cut and lay approx 17m of hedge, provide and erect a post and wire fence to the remainder to render the boundary stockproof
6	231/199	Bracken and briars encroaching into field. Eradicate and control to bring back into proper production
7	231/199 West end	Extensive growth of scrub and saplings. Cut and remove and provide proper fence

8	231/ Roadway	Weak, gappy, overgrown hedge. Cut and lay approx 50m of hedge, and provide post and wire stockproof fence to the remainder
9	231/ Roadway	Ditch overgrown and not cleaned out. Dig and clean out ditch to proper falls
10	231/ Roadway	Undergrowth, bracken and briars encroaching into field. Eradicate and control to bring back into proper production
11	231/ Roadway South of Lane	Weak, gappy hedge. Cut and lay and provide approx 43m of post and wire stockproof fencing
12	231/ Roadway	Undergrowth and bracken encroaching into field. Eradicate and control and bring back affected area into proper production
13	231/ Roadway	Barbed wire attached to oak tree. Remove
14	231/232	Encroachment of undergrowth, brambles, weeds and thistles into the field and at the boundary. Eradicate and control and bring back into proper production
15	231/232	Ditch overgrown and not cleaned out. Dig and clean out ditch to proper falls
16	232	This Pasture field is currently in Arable Production contrary to its scheduling on the tenancy agreement. Return to pasture and in doing so eradicate couch, argostis, buttercups, mayweed and also other arable weeds present
17	232/ Roadway West	Nettles, briars, undergrowth encroaching into the headland of this field. Eradicate and bring back into proper production
18	232/ Roadway West	Ditch overgrown and hedge weak and gappy. Dig out ditch and fence hedge line with post and wire stockproof fence

19	232/228	Ditch and surrounding area overgrown with undergrowth, briars, bracken and thistles. Dig out ditch, spread soil, cut, burn and eradicate undergrowth and bring whole area back into proper production
20	228	Stubble foul. Eradicate and control couch, argostis, docks and buttercups and other weeds
21	228/Lane	Encroachment of hedge and scrub woodlands into field. Cut and burn some 20m widths of thorn bushes, saplings, (mainly ash) dead wood, brambles, bracken, thistles, and nettles, eradicate and control to bring area back into proper production
22	228/Lane	Ditch overgrown. Dig and clean out to proper falls
23	228/227	Undergrowth, brambles, nettles, docks (approx 20m into field) encroaching into field. Cut. burn and control to bring back into proper production
24	228/234	High weak gappy hedge with blocked ditch either side. Cut and lay hedge making good the gappy area with stockproof post and wire fences, dig and clean ditches to proper falls
25	228/234	Undergrowth, briars, docks, nettles and fallen trees encroaching onto field, with weak unlayable hedge behind. Cut, burn and eradicate and control and bring area back into proper production. Provide post and wire stockproof fence on hedge line

3.   This Notice given in accordance with Case D in Part I of schedule 3 to the Agricultural Holdings Act 1986 and failure to comply with it within the period specified above may be relied on as a reason for a notice to quit under Case D.

4.   Your attention is drawn to the Notes following the signature to this Notice.

RIDER

1.   The above work to be carried out in a proper workmanlike and satisfactory manner.
2.   Hedges are to be laid in the proper season.
3.   Where ditches are cleaned out these must be cleaned to proper depths and falls and spoil to be properly spread.
4.   Where encroaching undergrowth to be cleared and eradicated all roots to be removed and area levelled as necessary.

5. Where hedges have been neglected and it is impossible to find suitable materials to lay the same, post and wire fencing (pig or sheep netting and one strand of barbed wire fixed on chestnut posts or tanalised posts) at 2m centres can be substituted, as indicated, in order to render boundary stockproof.

Dated the 31st day of October 2007

Signed   William Davies & Rees Agricultural Valuers
Address Dolanog,
         Welshpool,
         Powys.

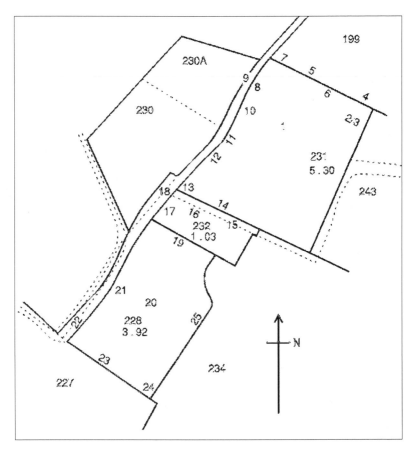

NOTES

In these Notes the Order means the Agricultural Holdings (Arbitration on Notices) Order 1987 (SI 1987 No 710)

What to do if you wish:

(a)   to contest your liability to do the work, or any part of the work, required by this Notice to remedy (Question (a)); or

(b)   to request the deletion from this Notice to remedy any item or part of an item of work on the ground that is unnecessary or unjustified (Question (b)); or

(c)   to request the substitution in the case of any item or part of an item of work of a different method or material for the method or material which this Notice to remedy would otherwise require to be followed or used (Question (c)).

1.   Questions (a), (b) and (c) mentioned in the heading to these Notes can be referred to arbitration under article 3(1) of the Order. To do so you must serve a notice in writing upon your landlord within one month of the service upon you of this Notice to remedy. The notice you serve upon your landlord should specify:

(a)   if you are referring Question (a), the items for which you deny liability

(b)   if you are referring Question (b), the items you wish to be deleted

(c)   if you are referring Question (c), the different methods or materials you wish to be substituted, and in each case should require the matter to be determined by arbitration under the Agricultural Holdings Act 1986 (c.5). You will not be able to refer Question (a), (b) or (c) to arbitration later, on receipt of a notice to quit. This action does not prevent you settling the matter in writing by agreement with your landlord.

Carrying out the work

2.   If you refer any of these Questions (a), (b) and (c) to arbitration, you are not obliged to carry out the work which is the subject of the reference to arbitration unless and until the arbitrator decides that you are liable to do it; but you MUST carry out any work which you are not referring to arbitration.

3.   If you are referring Question (a) to arbitration you may if you wish carry out any of the work which is the subject of that reference to arbitration without waiting for the arbitrator's award. If you do this and the arbitrator finds that you have carried out any such work which was not your liability, he will determine at the time he makes his award the reasonable cost of any such work which you have done and you will be entitled to recover this from your landlord (see article 8 of the Order). This provision does not apply in the case of work referred to arbitration under Question (b) or Question (c).

What to do if you wish to contest any other question arising under this Notice to Remedy

4. If you wish to contest any other question arising under this Notice other than Question (a), (b) or (c), you should refer the question to arbitration in either of the following ways, according to whether or not you are also at the same time referring Question (a), (b) or (c) to arbitration:

    (a) if you are referring Question (a), (b) or (c) to arbitration, then you must also refer to arbitration at the same time any other questions relevant to this Notice which you may wish to dispute. To do this, you should include in the Notice to your landlord referred to in Note 1 above a statement of the other questions which you require to be determined by arbitration under the Agricultural Holdings Act 1986 (see article 4(1) of the Order).

    (b) if you are not referring Question (a), (b) or (c) to arbitration, but wish to contest some other question arising under this Notice to remedy, you may refer that question to arbitration either now, on receipt of this Notice, or later, if you get a notice to quit. To refer the question to arbitration now, you should serve on your landlord WITHIN ONE MONTH after the service of this Notice to remedy a notice in writing setting out what it is you require to be determined by arbitration under the Agricultural Holdings Act 1986 (see article 4(2)(a) of the Order). Alternatively, you have one month after the service of the notice to quit within which you can serve on your landlord a notice in writing requiring the question to be determined by arbitration under the 1986 Act (see article 9 of the Order). You will then have three months from the date of service of that notice in which to appoint an arbitrator by agreement or (in default of such an agreement) to make an application under paragraph 1 of Schedule 11 to that Act for the appointment of an arbitrator. If this is not done your notice requiring arbitration ceases to be effective (see article 10 of that Order).

Warning

5. Notes 1 to 4 outline the only opportunities you have to challenge this Notice to remedy.

Extensions of time allowed for complying with this Notice to remedy

6. If you refer to arbitration now any question arising from this Notice to remedy, the time allowed for complying with the Notice will be extended until the termination of the arbitration. If the arbitrator decides that you are liable to do any of the work specified in this Notice to remedy, he will extend the time in which the work is to be done by such period as he thinks fit (see article 6(2) of the Order).

Warning as to the effect which any extension of the time allowed for complying with this Notice to remedy may have upon a subsequent notice to quit

7.   If your time for doing the work is extended as mentioned in Note 6 above, the arbitrator can specify a date for the termination of your tenancy should you fail to complete the work you are liable to do within the extended time. Then, if you did fail to complete that work within the extended time, your landlord could serve a notice to quit upon you expiring on the date which the arbitrator has specified and the notice would be valid even though that date might be less than 12 months after the next term date, and might not expire on a term date. The arbitrator cannot, however, specify a termination date which is less than 6 months after the expiry of the extended time to do the work. Nor can he specify a date which is earlier than would have been possible if you had not required arbitration on this Notice to remedy and had failed to do the work (see article 7 of the Order).

Example Tenant's Notice requiring Arbitration

AGRICULTURAL HOLDINGS ACT 1986 SCHEDULE 3, PART 1, CASE D

AGRICULTURAL HOLDINGS (ARBITRATION ON NOTICES) ORDER 1987
(SI 1987 No 710)

Re: The holding known as Part of Lower Barn Farm, Great Onen, Monmouth, Gwent — OS Nos 231, 232 and 228

To: Mr J.H. Fitzherbert, Landlord
c/o his agents — Williams, Davies & Rees
Dolanog
Welshpool, Powys

I HEREBY GIVE YOU NOTICE as follows:
I wish to contest my liability under the terms or conditions of my tenancy to do the work specified in the Notice to do work dated 31 October 2007 which you have given me in respect of the following items:

1	OS 231	Pasture Foul. Eradicate couch, agrostis, docks, thistles, chickweed to bring field back into full production.
2	OS 231/ 243	Nettles, bracken and briars encroaching into field. Eradicate and control to bring area affected back into proper production.

3	OS 231/ 243	Weak Gappy Hedge. Cut and lay and render stockproof.
4	OS 231/ 199	Ditch overgrown and not cleaned out. Remove overgrowth, shrubs and briars and clean out to proper falls.
5	OS 231/ 199	Weak, gappy, overgrown hedge. Cut and lay approx 17m of hedge, provide and erect a post and wire fence to the remainder to render the boundary stockproof.
8	OS 231/ Roadway	Weak, gappy, overgrown, hedge. Cut and lay approx 50m of hedge, and provide post and wire stockproof fence to the remainder.
9	OS 231/ Roadway	Ditch overdrawn and not cleaned out. Dig and clean out ditch to proper falls.
13	OS 231/ Roadway	Barbed wire attached to Oak Tree. Remove.
16	OS 232	This pasture field is currently in Arable Production contrary to its scheduling on the tenancy agreement. Return to pasture and in doing so eradicate couch, agrostis, buttercups, mayweed and also other arable weeds present.
20	OS 228	Stubble Foul. Eradicate and control couch, agrostis, docks and buttercups and other weeds.
21	OS 228/ Lane	Encroachment of Hedge and Scrub Woodland into field. Cut and burn some 20m width of thorn bushes. Saplings (mainly ash), deadwood, brambles, bracken, thistles and nettles, eradicate and control to bring area back into proper production.

The grounds on which I deny liability to do such work in respect of the aforesaid are as follows:

Item No 1   I deny that this pasture is foul and infested with the weeds alleged any more than expected of an old pasture of this age and kind. The proper solution would be to plough up and re-seed but I am prohibited by Clause 10(4) of my tenancy agreement from doing this.

Item No 2   I deny that the alleged infestation exists. There is no loss of production. It is practically impossible to eradicate the type of infestation alleged, if proved that it does in fact exist.

Item No 3   This hedge does not belong to the property I rent. This has always been maintained by the neighbour Mr AN Other.

Item No 4   This ditch is not my responsibility and belongs away. In any event this ditch does not require cleaning out. The sides have been recently cut back by me and I always do this every year, in my own interest, but without accepting any liability to do so.

Item No 5   This boundary belongs away and has never been maintained by me during my tenancy.

Item No 8   This hedge is not ready for laying. It has been left to grow for laying in about 4 years' time when it will be of sufficient growth. The boundary is stockproof and there is no need to provide the post and wire fence requested. Furthermore, I have absolute discretion as to which hedge I cut and lay each year. I am only required to cut and lay a proper proportion and this has been done.

Item No 9   This ditch does not require cleaning out. It was cleaned out last year by a specialist contractor. The sides have recently been trimmed.

Item No 13   This barbed wire was not attached by me to the oak tree and existed as such when I took the tenancy.

Item No 16   Permission was granted by the previous owner to plough this field (copy letter dated 16 August 1968 attached) I am not required to re-seed the same. It is denied that a serious infestation of weeds exist, particularly couch. The other weeds will be dealt with at the appropriate time in carrying out normal husbandry.

Item No 20   It is denied that the stubble is foul except that a minor infestation of annual weeds exists and these will be dealt with at the appropriate time.

Item No 21   I have no liability for reclaiming this area. It has always existed in this state and was so at the commencement of my tenancy. I have an old OS Map dated 1930 which shows it as rough.

I request all the above items to be deleted from the notice.

I also wish to contest the following questions on the Notice to do work dated 31 October 2007 which you have given to me in respect of the following matters:

(a) my liability, if any, to do the work stipulated and also the manner in which you prescribe it should be done
(b) the enforceability of the Items 1–5 of the Rider forming part of your Notice
(c) as to whether the period of 12 months given to carry out the work is a reasonable period. In my view it is not and is far too short a period in which to carry out all, or any, of the work stipulated.

I require the aforesaid questions and matters to be determined by Arbitration under the 1986 Act.

Dated this 10th day of November 2007

Signed: A Charles (Tenant)
         Great Onen Farm
         Monmouth

Practice Note
Although the example reply does not show this, it is prudent when giving a notice requiring arbitration to oppose EVERY item. If this is done, items that are agreed can be conceded at the hearing (if the tenant has not in the meantime done them) and the arbitrator can include the same in his award or if the work is done, the landlord can withdraw on those particular items. If arbitration is not asked for on every item the work stipulated must be done.

Care should also be exercised in the preparation of the notice to remedy not to include items which are the responsibility of the landlord to repair.

Probate Valuations

15.1 General

Agricultural probate valuations usually cover both freehold agricultural properties, live and dead farming stock, and also quotas, where the deceased owned the same at the date of death.

15.2 Freehold properties

The valuation should be as comprehensive as possible and fully describe the property in detail, complete with a schedule of OS numbers and areas, and have full particulars of outgoing charges. If possible a plan should be attached to the valuation. All separate parts should be adequately described including cottages, buildings, etc. If the farm is let, full details of the tenancy, term date, rental reserved (and when last revised) and repairing and insuring liabilities should be given. Also, if the tenants or occupiers (where they are not the same as the deceased) have carried out improvements (whether by consent or not) these should be carefully listed, as the value of these improvements are to be disregarded since they do not belong to the deceased's estate (if valuing for the landowner's estate).

If the property is in joint ownership, a 10% deduction should be made after dividing the share (if in the ownership of more than two persons, up to 15% can usually be deducted).

Consideration should be given in valuing the farmhouse. In the *Antrobus* case (Worcestershire) it was ruled that the agricultural value of a farmhouse is less than its open market value — possibly a low as

70% which will leave up to 30% of the value of the house chargeable to inheritance tax. This is possibly particularly applicable to small farms or high quality farmhouses which form an undue proportion of the value of the farm as a whole, eg a farm of say 53ha worth as a whole say £1,300,000 but where the value of the farmhouse and gardens is itself worth say £1,000,000 leaving £300,000 as the value for the remaining land and buildings. The *McKenna* case in Cornwall further reinforced this law. It was stated by the special commissioner that 'The farmhouse is a dwelling for the farmer from which the farm is managed and that the farmer of the land is the person who farms it on a day-to-day basis.'

In both these cases the farmhouse was of high quality and in the *McKenna* case neither the owner nor his widow occupied the house for the purposes of agriculture in the two years leading to his death.

15.3 Live and dead farming stock

Here, it is essential that everything should be described properly and individual values placed on each animal and item of produce, crops and dead farming stock. All animals should be described by their breed, age, etc, and valued accordingly. Dairy cows should be valued individually, with names or numbers given as should be rearing cows with calving and service dates given. Store cattle and sheep can be valued in matching bunches. It is always important, to give, where possible, the ages of animals valued. A separate value should be given on each item of machinery, tractors, tools, etc. In the case of important and valuable equipment, the year of manufacture or registration (if new, then) should be given.

This is not work that should be skimped, but should be dealt with great care and in detail, as otherwise endless queries will arise from the Capital Taxes Office.

15.4 Reliefs

In forwarding the valuation to the deceased's solicitor, it is prudent to remind him that where the requirements are in met and the property is owner-occupied, 100% and 50% relief when the property is let should be claimed. Also, 100% Business Asset relief should be claimed from the value of the live or dead stock.

Example of probate valuation

VALUATION FOR PROBATE
In the Estate of
FREDERICK ARNOLD Deceased
late of
BANK FARM, SCOTFORD, HEREFORD
Date of Death: 16th day of December, 2007
Date of Valuation: 10th day of January, 2008

(A) **FREEHOLD PROPERTY**
BANK FARM, SCOTFORD, in the County of Hereford & Worcester
An agricultural investment comprising Farmhouse, a detached service cottage, farmbuildings and pasture and arable land, together with a small area of woodland, extending to 84.52ha.

The holding is situated adjoining a council maintained road on the Scotford to Pearton road and is about 5 miles from the City of Hereford.

(a) **The Homestead**
The Farmhouse is constructed of brick, stone, part half timber and has a slate and clay tile roof. It is thought to date from the mid-eighteenth century. It contains:

On the Ground Floor:
Entrance Hall
Two Reception Rooms
Dining Room
Kitchen with Aga Cooker and Boiler
Scullery
Dairy Washhouse

On the First Floor:
Landing
Five Bedrooms — 3 double; 2 single
Bathroom: with heated linen cupboard
Separate WC
Boxroom

On the Second Floor:
Landing
Three Attic Rooms
This floor is disused

Outside: Large garden with small orchard

The Farmbuildings
These are convenient to the residence and comprise:
Steel framed concrete block and asbestos Loose Housing
Covered Cattle Yard with raised centre feeding passage.
Dairy Complex: (now disused) comprising 8/16 Herringbone Milking Parlour
with Milk Room adjacent.
Circular concrete Collecting Yard.
Steel, asbestos and concrete Silage Barn.
Traditional stone built and slate roofed Barn with 5-bay-open-sided cattle
shed fronting an open yard.
Brick and asbestos Implement Shed/Garage.

(b) **The Cottage**

This is a detached brick and slated service cottage known as Stopforth
Cottage and is situated off the entrance drive. It is on two floors and
contains:

Hall
2 Living Rooms
Kitchen
Scullery

On the First Floor:
Landing
Three Bedrooms
Bathroom

Outside:
Garage and Garden.

(c) **The Land**

This is in a ring fence and is mainly level or gently undulating. The soil is a
light, free draining loam, derived from the old red sandstone formation and
the land comprises 10 enclosures, together with a small coppice of 4H
acres of young ash, oak and pulpwood. The land is watered with a number
of automatic water troughs from the mains supply and also an intersecting
stream. The whole extends to 84.52ha as contained in the following
Schedule taken from Herefordshire 1/2500 OS Sheet L10/11.

OS No	Description	Area (Acres)
782	Homestead	1.01
654	Pasture	24.65
821	Pasture	20.10
863	Pasture	18.76
158	Arable	15.19

242	Arable	12.92
111	Pasture	27.98
515	Pasture	24.81
992	Coppice	4.56
002	Pasture	18.76
004	Roadway	1.04
182	Pasture	21.11
259	Pasture	17.98

	208.87	acres
	(84.52	ha)

(d) **Tenancy**

The holding is let on an annual Candlemas agricultural tenancy to Mr Fraser-Wood by virtue of an agreement dated 10 February 1986, the tenancy having commenced on 2 February 1985. Repairing liabilities are in accordance with SI 1973 No 1473. The rental currently payable, with effect from 2 February 2007 is £10,000 per annum.

The tenant has a son, working full time with him, aged 28 years.

(e) **Tenant's Improvements**

The tenant, at his own expense, has carried out the following improvements.

(i)   Modernised the farmhouse including provision of hot water system, bathroom (complete with all fittings), installed Aga Cooker and provided all kitchen equipment.

(ii) Constructed the farm drive including finishing with an asphalt surface.

(iii) Tile drained OS Nos 821 & 182 (36 acres).

(iv) Installed the whole of the dairy complex comprising Herringbone Parlour, Milk Room, Collecting Yard, Covered Cattle Yard (90' × 60') and the Silage Barn.

Landlord's written consent for all this work has been given, depreciating the cost of each item down to £1 over a period of 15 years.

(f) **Outgoings**

Drainage Rate	£58.70
Charity Rent	£10.25

(g) **Quotas**

The holding has a wholesale milk quota of 292,061 litres, of which, it is considered, informally, the deceased owned 70%.

(h) **Services**

Mains electricity and water connected.
Drainage of residence and cottage to individual septic tanks.

(i) **Valuation**

Subject to the existing tenancy and including 70% of the milk quota, but disregarding the value of the tenant's improvements or his SFP £416,000 (four hundred and sixteen thousand pounds).

(B) **LIVE & DEAD FARMING STOCK** at MANOR FARM, SCOTFORD (example when deceased was farming the holding)

(a) **Cattle**

41	20 month-old Steers:		
	25 Hereford cross Friesians @ 480	£12,000	
	15 Charolais cross @ £510	7,650	
	1 Shorthorn cross	420	
20	Hereford cross Friesian		
	15 month Steers @ £360	7,200	
61	Total		27,270

(b) **Sheep**

184	Breeding Ewes:		
	105 Suffolk cross 3–4 year old @ £38	3,990	
	61 Mule 2-year-old @ £60	3,660	
	18 Mule Stock Ewes (BM) @ £32	576	
	2 Suffolk 2-year-old Rams @ £80	160	
186	Total		8,386

(c) **Pigs**

10	Large White Cross Landrace Sows with 91 piglets (up to 6 weeks old) @ £240	2,400	
4	Large White Cross Landrace Sows due to farrow within 1 month @ £150	600	
10	Large White Cross Landrace Sows served 1 month to 2.5 months @ £125	1,250	
5	Large White Cross Landrace Sows recently weaned @ £100	500	
1	Large White Stock Boar	200	
80	Pork weight pigs @ £55	4,400	
70	Medium Store pigs @ £35	2,450	
35	Weaners @ £22.00	770	
306	Total		12,570

(d) **Agricultural Equipment**

1979 Commer Stocklorry with 24' Box Reg No AHN 007M 121,600 recorded miles	£600
1995 MF 595 Tractor Reg No HGV 116	4,100
1982 MF 135 Tractor Reg No SHJ 765	1,000
1987 187 SP Combine Harvester 10' cut Reg No RTM 112R	1,500
Urry Cattle crush	140
Circular Cattle Feeder	40
6–8' metal cattle racks @ £35	210
Wooden Cattle Rack	25
4 Covered Sheep Racks @ £50	200
1 Covered Sheep Rack (very poor order)	10
10 Sheep Troughs 8 @ £10	80
2 (very poor) @ £2	4
Fraser 8 Grain Trailer	1,500
FR 5 Furrow Plough	500
Fahr KM 22 Mower	250
Hay Bob	150
MF 3 trailer	200
IH Fore-end Loader with Fork, bucket and bale spike	525
Parmiter Silage Grab	400
Scrap Iron	10
Small Tools	30
B & D Electric Drill	25
MF 30 G & F Drill (3 m)	850
IH 440 Baler	400
Howard 150 Rota Spreader	450
Vicon 3 m Power Harrow	1,200
Cambridge Gang Rollers (18')	425
Parmiter 13' zig-zag Harrows	200

15,024.00

(e) **Produce (Consuming Values)**

Hay	
5,500 bales 2006 baled seeds hay (very weathered) = 110 tonnes @ £40	4,400
225 tonnes Silage @ £12	2,700
25 tonnes 2006 baled Wheat Straw @ £15	375

7,475.00

(f) **Growing Crops**
   OS Nos 104 & 2002 — 8.09ha Avalon Winter
   Wheat @ £183.43ha (cost of seeds,
   fertilisers, cultivations and enhancement value)   1,484.60

(g) **Tenant's Pastures**
   OS No 12 6.92ha (4 yr ley sown 1998)
   6.92ha @ £32.51   225
   Pt OS No 87 8.09ha (LT ley sown 2000)
   @ £182.85   800
                                                          1,055.00

NOTE: No valuation has been made of any remaining improvements (short and long term) or unexhausted or residual manurial values since it is considered that these will offset a dilapidations claim that will arise on the termination of the tenancy.

SUMMARY

(a)	Cattle	27,270.00
(b)	Sheep	8,386.00
(c)	Pigs	12,570.00
(d)	Agricultural Equipment	15,024.00
(e)	Produce	7,475.00
(f)	Growing Crops	1,484.60
(g)	Tenants Pasture	1,055.00
	Total	£73,264.60

GRAND SUMMARY

Freehold propety	£416,000.00
Live and dead farming stock	73,264.60
	£489,264.60

We, the undersigned, having inspected the above, DO VALUE, the same, at the date of death, in the figures given.

. . . . FRICS, FAAV
of Messrs Williams, Davies & Rees
Dolanog
Welshpool
Powys

Stock Taking or Income Tax Valuation

16.1 These valuations generally form an important part of any agricultural valuation practice. The work, system wise, is repetitive and most valuers use special valuation pro formas for completion with numbers and the detailed calculations. Adopting this system means that items are not forgotten and the valuation is done in a standard and consistent way each year: See example at 16.6.

16.2 The Inland Revenue have issued guidance notes — Business Economic Note 19 — setting out the basis of agricultural stocktaking valuations. This basis should be carefully followed particularly since the introduction of self-assessment.

The basic principle of BEN 19 is that stocktaking should be based on the lower of cost or net realisable value. This is usually a cost of production basis, ie the actual costs, but the deemed cost method is acceptable where actual costs are not known and stock is then valued on the percentages given:

Cattle	60% of open market value
Sheep Pigs	75% of open market value
Harvested crops	75% of open market value

Harvested crops include hay, silage, straw (produced on the farm and not purchased). Growing crops should be valued on a cost of seed, cultivation, fertiliser and spraying basis, any fertiliser applied to pasture (and not grazed or crop taken) and for acts of husbandry such as orchard spraying should be included at cost. Enhancement value is not included. CAAV are expressly recognised in para 3.1.5.1 of BEN 19 as the basis of establishing actual costs of production.

Directions made by HM Revenue and Customs are given to the farmers accountant to include the single farm payment payable in respect of any particular holding in the annual accounts, but this is purely an accountancy matter and does not concern the valuer.

16.3 In the case of livestock purchased and reared on, the valuation should be the higher of the cost and 60% of market value of cattle and 75% of the value of all other livestock. Check that the cost does not exceed the market value and value on the lower figure.

Harvested crops such as corn, potatoes, sugar beet, etc, should be valued at lower of cost and net realisable value. Calculate cost of production to determine cost. Many farmers do not keep records and where this is so, it has been agreed between the Revenue and the NFU that 75% of the market value on the holding on the date of valuation can be adopted.

Hay, straw and silage, produced on the farm, should be valued at cost of production or at 75% of market value where no cost figures are available. Purchased fodder should be valued at the lower of cost and net value.

In the case of growing crops these are valued at cost of seeds, cultivations, sprays and fertilisers. Enhancement value is not included. Any fertilisers applied to pastures and other acts of husbandry such as orchard spraying should be included at cost.

16.4 Note, where the production herds (this only applies to milking and rearing cows, bulls, breeding ewes, rams, sows and boars) are on the herd basis no value should be given for these but the actual numbers should always be shown on the valuation. Note, the herd basis is only applicable to production stock and not to first calving heifers, in-pig gilts, or in-lamb hoggetts which at the date of valuation have not done any producing. If the ewe flock has lambed, and the flock is on the herd basis, only the lambs are valued and not the ewes. Similarly, where there is a suckler herd, again on the herd basis, only the calves are valued.

16.5 It should be borne in mind that this is a stocktaking valuation comprising the livestock, fodder, produce, stores, growing crops, acts of husbandry and fertilisers applied on a farm. It does not embrace equipment, fencing, and drainage works. Very few valuers include unexhausted manurial values, unexhausted lime and residual value of feeding stuffs. These are not usually charged for when an owner occupier vacates a farm and where a tenant vacates — he might well have a dilapidations claim against him and these items, in a tenant's case, would serve as a partial offset against such a claim.

16.6 Guidance notes on this type of valuation are published by the CAAV and RICS as numbered publication No 170.

Example of valuation pro forma

INCOME TAX VALUATION
NAME...................... VALUATION AS AT.................
ADDRESS.................. DATE OF VALUATION ....
....

CATTLE				Details	@	£	p	£	p
Milking Cows	....	....	....						
Rearing Cows	....	....	....						
Bull(s)	....	....	....						
....									
In-calf Heifers	....	....	....						
Two-Year-Old	....	....	....						
18/20 months	....	....	....						
12/15 months	....	....	....						
6/10 months	....	....	....						
2/5 months	....	....	....						
Young Calves	....	....	....						
....									
SHEEP									
Ewes in/and with lambs viz:									
....									
Rams	....	....	....						
Lambs	....	....	....						
Cull Ewes	....	....	....						
Wethers	....	....	....						

HORSES	Details	@	Forward			
			£	p	£	p
. . . .						
PIGS						
In-farrow Sows						
Breeding Sows						
Breeding Sows and Litter(s)						
Boar(s)						
In-farrow Gilt(s)						
Breeding Gilt(s)						
Weaners						
Porkers						
Baconers						
. . . .						
POULTRY						
Hens						
Chicks						
Ducks						
Turkeys						
FEEDING STUFFS/CORN IN STORE						
T						
T						
T						
HAY						
T						
STRAW						
T Barley Straw						
T Wheat Straw						
T Oat Straw						
SILAGE						
T						
ROOTS						
T Mangolds						
SEEDS & SEED CORN						

	Details	@	Forward £	p	£	p
SEED POTATOES						
T						
T						
WARE POTATOES						
T						
FERTILISERS IN STOCK						
T						
T						
T						
AGRICULTURAL STORES						
. . . .						
. . . .						
FUEL						
L						
L						
GROWING CROPS & ACTS						
OF HUSBANDRY						
Ha Wheat						
Ha Oats						
Ha Barley						
Ha						
. . . .						
Ha						
. . . .						
Ha Potatoes						
Ha Swedes						
Ha Kale						
Ha Sugar Beet						
Ha Young Seeds Planted						
. . . .						
. . . .						
Ha Manures & Lime Applied:						
. . . .						
. . . .						

William Davies & Rees
AGRICULTURAL VALUERS
Dolanog, Welshpool, Powys

Example income tax valuation

FAIR OAK FARM
HENDLEFORD
INVENTORY AND VALUATION

of the Live and Dead Farming Stock, Produce, Fertilisers, Stores, Fuel, Growing
Crops and Acts of Husbandry

the property of MJL Yeoman
as at 25th March 2007

CATTLE:

78	Milking Cows	} HERD BASIS	
6	Rearing Cows		
1	Bull		
74	Other Cattle		£20,560.00

SHEEP

172	Ewes	} FLOCK BASIS	
4	Rams		
299	Other Sheep		3,837.00

FEEDING STUFFS	3,864.00
HAY	2,130.00
STRAW	672.00
SILAGE	4,180.00
FERTILISERS IN STOCK	1,955.00
AGRICULTURAL STORES	197.60
FUEL	208.00
GROWING CROPS & ACTS OF HUSBANDRY	9,717.20
	£47,320.80

NOTE: Single farm payment are not included in this valuation.

. . . .   MRICS, FAAV
of Williams, Davies & Rees
Chartered Surveyors and Valuers, Dolanog, Welshpool, Powys

Valuation of Freehold Farms

17.1 General

Valuation of agricultural freeholds and their rental value is a skill only acquired after a number of years' experience, and the young valuer will have an advantage if he has actually worked on a farm and will learn of the difficulties or advantages of any particular farm. There are many factors to consider including those set out below.

17.2 Quality of the soil

The productivity and thus profitability of any farm (except an intensive production unit such as a pig or poultry farm) depends on the quality of the soil. The ideal is a fairly easily worked soil, preferably deep, free-draining level loam (but not too sandy as it dries out). It should be versatile to the extent that almost any crop can be grown on it, and should be ideal for all arable crops as well as livestock enterprises dependent on grass, such as dairying.

Frequently, much level land is clay based and thus heavy and often water-logged. Potatoes and sugar beet cannot be satisfactorily grown on heavy land because they cannot always be harvested in the wet conditions experienced in the autumn. Similarly, it may sometimes prove difficult to plant such land with autumn sown cereals. This type of soil usually has a high water table, but in dry conditions tends to dry out very quickly and large cracks develop. Examples of this soil are to be found extensively in Worcestershire and also the Severn Vale in Gloucestershire. This soil is also cold and late in the spring.

An ideal soil is often an alluvial loam of good depth with little stone content. This is generally relatively free-draining, does not get water-logged and can grow most crops. However, this type of soil is normally found in valley bottoms, some of which are prone to flooding.

Sandy soils, while easy to work, dry out very quickly and unless the growing season and the summers are wet, yields suffer by comparison with crops grown on rich loamy soils. Sandy soils also wear out ploughs and cultivating equipment and tyres because they are so abrasive. However, these soils have the advantage of being early in the spring and are also good for late summer and autumn growth. Generally, it should be possible to harvest sugar beet and potatoes, and also to plant autumn sown crops when there may be difficulty on heavier soils. Sandy soils benefit considerably where irrigation facilities exist. They also tend to blow in certain situations, particularly in the eastern counties where there are few windbreaks and established vegetation. The yields are seldom as good on light soils, as on heavier soils. In particular, pastures do not grow as thick and grass plant population is usually less. Sandy soils are often hungry soils and frequently are deficient in lime, phosphates and potash which tend to leach easily, particularly on banky fields. The soil on banky fields also tends to wash out easily during heavy rainstorms.

Many soils, particularly at higher elevations in limestone areas, are very stony and rocky. Often there is little depth of soil and the underlying rock formation is near the surface. Crops suffer in drought periods and sometimes completely die off or do not develop. These shallow soils are usually only suitable for grazing but, nevertheless in the last three decades many have been ploughed. Cereals are often the only arable crops that can be satisfactorily grown, since they have so little depth (often only 12cm or less top soil). High value cash crops, such as potatoes and sugar beet, cannot be satisfactorily grown as these need deep soils, and the presence of stone and rock does not allow mechanical harvesters to function properly. Stony and rocky soils burn in periods of drought and are also very damaging to tractor tyres, and implement breakages and damage are frequent. Examples of such soils are to be found in the Forest of Dean, the Cotswolds, Pennines and the flinty areas of the southern counties. The existence of stone boundary walls is the first indicator of the presence of stone, apart from the rock outcrops which sometimes are visible. Trees are usually stunted and hedges are sparse and often are only a few clumps here and there.

Sometimes clay patches, which are heavy and wet, and also areas with rising springs, causing wet patches, are found in all classes of land, even in sandstone soils. These as a rule need draining.

The presence of thick continuous hedgerows, well grown, and also good large trees is usually an indicator that the soil is deep and of good quality. The presence of elm trees (although now devastated by Dutch Elm disease, but elm hedges still exist) is an indication of good, deep, rich soils.

Ditches and watercourses are seldom found on free-draining soils, particularly in sandy soils, chalk soils and limestone areas. The exception is in the valleys and river flood plains where the soil tends to be heavier, and is often clayey loam. The absence of ditches can be an advantage, since they do not have to be cleaned out, but may be also a disadvantage if there is no other water supply available for stock.

Some soils overlie impervious clay strata and unless drained are perpetually water-logged, unproductive and grow, in particular, rushes and sometimes reeds. Drainage is not always effective, especially if the area is very low lying. Contrast the Fenlands, which were made into some of the most productive arable land in this country by drainage despite, in many areas, being below sea level. Very deep drainage channels and levels (Bedford Levels) were dug and the water pumped out into the sea. The Fens are of a peat soil which is inherently rich and frequently capable of producing two crops a year. Here, in this area of comparatively low rainfall, the presence of a fairly high water table is an asset as is irrigation.

17.3 The homestead

The existence of a good quality residence has an important bearing on farm values, now that the values of residences has escalated. Very often, when farms are sold, the residence gardens and a small paddock is lotted separately and is often purchased by a non-farming buyer.

It is very advantageous to have good modern buildings, particularly well designed, soundly built and properly maintained. Specialist buildings such as poultry and pig units do not appeal to every type of buyer and, if extensive, can be a hindrance to a sale, particularly if there is a slump, say, in egg or broiler profitability or the pig industry, as occasionally occurs.

Dairy buildings, especially if well designed with a milking parlour, cubicle housing and silage pits, on smaller farms (say up to

80ha) are an asset to a sale, since many farmers who farm holdings of this size are dairy farmers. If they did not exist it is a costly matter to provide a modern dairy set-up. A milk quota is, of course, essential to practise dairying but purchasers of holdings without quota can purchase quota if they intend to practise dairying and where no, or insufficient quota, exists.

The most popular and useful buildings are portal-framed general purpose buildings which can be used for housing livestock, whether beef or dairy cattle, used for sheep housing, storage of grain, potatoes, hay and straw or farm equipment. Such buildings should have high headroom (minimum 4.87m at eaves), (low headroom buildings are unhealthy for cattle as they frequently get virus pneumonia attacks) and be maintenance free — eg precast concrete and galvanised steel which does not rust and does not require painting. Also, if the building is a covered silage pit it must have a high headroom, as tractors with safety cabs cannot otherwise have enough headroom to operate when consolidating the silage.

The existence of good modern grain stores with drying facilities, which are vermin- and bird-proof is an obvious asset, as is the existence of refrigerated potato stores where potatoes are grown.

The presence of adequate farm cottages on larger holdings, where they are needed, is a further asset.

If the farmbuildings on any given holding are old, traditional buildings (and with no modern buildings), this will, in some cases, result in the capital value being a good deal less than when a farm is suitably equipped with modern buildings. However, old stone buildings which can be converted into dwellings or holiday lets, provided they are a suitable distance from the residence, are nowadays an important plus factor, in the value of any holding.

Many farms equipped with modern buildings have been badly designed, are of a poor layout, and sometimes such buildings are too specialist in nature, and in some cases these buildings are now out-dated. Examples are: buildings which have too low a headroom, cattle buildings with badly designed feeding and bedding arrangements (cattle cannot be fed by tractor and forage wagon, and cattle cannot easily get access to silage clamps); slatted floor buildings (not every farmer wants to keep his cattle on slats and the buildings really cannot be used for anything else); cattle buildings with tower silos and automatic feeding of tower silage systems (not everyone wants to be dependent on this system as if there is a power cut the cattle cannot be fed, apart from the numerous mechanical breakdowns that occur);

concrete floors not damp-proofed and where grain cannot be stored owing to rising dampness.

It follows, therefore, that where a holding is inadequately equipped, if it can be purchased at a lower figure to reflect the cost of essential buildings, this can be beneficial since exactly what is needed can be erected.

17.4 Situation

The location of a farm has a major bearing on values and the following matters are significant.

17.4.1 Position in relation to towns

A holding convenient for towns and markets, with shops, schools and other facilities, is usually worth more than a holding that is remote. Transport costs time and money. Conversely, if a farm is on the outskirts of a town, this can be a nuisance because of trespass and vandalism, and often a sheep flock cannot be kept owing to dogs worrying sheep. Also, a number of footpaths usually exist on holdings near towns and these are often much used and result in further trespass.

17.4.2 Position in relation to roads

A farm with the homestead immediately adjacent to and with direct access from a much used road can be a dangerous place to live. Fast-moving traffic is a major hazard to the movement of tractors and equipment and also livestock. Where some land exists across a road and a dairy herd is kept, the movement of cows four times a day for six months of the year across this road is a positive hazard. Vehicles run into stock and kill or injure them, and every time stock is taken across, two or three persons are needed to warn traffic and drive the stock.

Conversely, a farm situated with its homestead some distance, say up to 5–600 metres from a main road and with no or very little land severed, has many advantages.

It is not always an advantage to have a farm approached by a long narrow council road as there are frequently no proper passing places, and vehicles have to proceed with caution. However, a little used road can sometimes be an asset, since it might give access to roadside fields and is, of course, particularly useful in wet weather to give firm access to these fields. Farms with their own long private drives leading off roads have the disadvantage that the drive is costly to maintain.

In this age of reduced farm profits a situation where a farm shop can be operated, buildings converted for alternative use, or a Caravan Club Licence obtained can be a major asset as these can be profitable diversifications.

17.4.3 Farm approaches

A farm with a good level access has a distinct advantage over a holding approached up a steep hill, which can be difficult in winter. Likewise, if the access road is narrow and exposed it can easily become blocked with snow drifts. Occasionally, farms have low-lying accesses which are liable to flooding, and this can be a problem at times.

17.4.4 Farms on flood plains

Valley land is usually good productive land, but if it is liable to flooding this can be a drawback since, in some valleys, the river floods on a number of occasions annually, mainly in the winter. This restricts its user and losses can arise to cereal crops, and it is frequently unsafe to grow potatoes and root crops, in particular sugar beet, since they cannot always be harvested. The odd summer deluge resulting in summer flooding can result in serious losses to growing crops. If a farm homestead is itself on a flood plain, and is susceptible to flooding, this can be a serious problem. Land on flood plains usually has a large number of ditches and water courses have to be cleaned out, and this can prove an expensive, recurring liability.

17.4.5 Elevation and aspect

A holding at a low elevation is generally more productive and often has a great deal more inherent fertility than a farm at a high elevation, of say 160m and upwards. The growing season is longer and the farm is generally much earlier than a high-lying farm, which is often (unless in a sheltered high valley) cold and late. However, low-lying farms in the eastern counties, being on wide open plains, tend to be cold and exposed. The growing season is also longer in the south of the country than further north, but the earliness of any holding usually depends on whether it has good free-draining sheltered soil, and also its location. The influence of the Gulf Stream is an important factor and in western areas, eg Cornwall, Pembroke etc, early potatoes are grown since some of these areas are frost-free. Much low-lying land is, however, wet and the soil is of clay and impervious and has a high water table, and is consequently late in spring eg the Somerset Wetlands. It is sometimes difficult to drain this wet land, particularly if very low lying, and drains, when laid, frequently get blocked up. The result is rush-infested, unproductive land, which can only be grazed for a limited period in the summer months.

The aspect of land is very important. If it faces north east, it is cold and late. Snow and frost clears last of all from land with these aspects, by contrast to land with a south and west facing aspect having plenty of sun. However, in very hot summers, north and east facing land does not burn to the same extent, particularly where rock is near the surface. Land adjacent to woodlands, especially if it is on a north and east slope, is usually very unproductive, and it is rare for anything much to grow close to woodlands and under trees. Also, hedges do not grow well by woodland, and it is to be noted that rarely is a good hedge to be found under a tree or trees. Overhanging boughs can also be a nuisance in the use of machinery, particularly high tractors and combine harvesters.

Banky land is difficult land to farm. If the banks are severe, tractors with twin wheels (and four-wheel drive) or caterpillars are needed for reclamation, re-seeding, fertilising and thistle-cutting. However, all banks are dangerous, and considerable caution must be exercised with the use of any tractor and vehicle. Steep land is not generally suited for arable production and combine harvesters do not function well on banks as some of the grain harvested is discharged through the back of the machine. This type of land is generally most suited for grazing purposes. Banky fields have a greater surface area than that shown on the OS maps! This is why many marginal land and hill farms carry greater headage of stock than one would expect them to keep. Adequate rainfall is extremely important.

17.5 Fencing, gates, ditches and drains

A most important factor on any stock farm is the presence of proper stockproof hedges or fences.

17.5.1 Hedges

Good stockproof hedges are essential on stock farms. They serve to keep stock in the fields and also give shade from the sun and protection from wind and rain. This is why so many high hedges, which one may consider should be laid, exist on stock farms. It is also wise to allow a few hedgerow trees to grow on instead of trimming them as they provide shade and shelter.

Hedges grow well on rich loamy and heavy soils but do not thrive on thin soils. A hedge usually takes 10–12 years from the date of planting to the first laying. If a new hedge is planted, it should be of quickthorn, planted in two rows, and staggered.

17.5.2 Fences

The most suitable economic fence on a farm is medium/heavy gauge pig netting, erected on tanalised posts (erected every 2m with adequate straining posts and struts at the end or at any change of direction). Two strands of barbed wire should be provided to provide a minimum height of 1.22m. Such fence should, if properly erected, last for 15–20 years, unless situated in an area of industrial pollution or near the seaside.

When holdings are converted to total arable farms, hedges and fences are often neglected and the only attention hedges have, as a rule, is trimming. If the farm is at a subsequent time used as a livestock holding, it is a very costly and time-consuming matter to erect permanent stockproof fences.

17.5.3 Gates

Are important on any mixed or stock farm. A well-run farm should have good quality iron gates of adequate width (preferably 4.57m wide or double gates of 6.09m width) properly hung and with proper fasteners (not fastened with baler twine!). It is important that a gate, when erected, should have adequate strength and not be too light in construction. Cattle rubbing against a light gate will soon buckle it and will bend the top rail if they try to jump over when it has been hung too low.

17.5.4 Ditches and drains

Where these exist, they should always be maintained and kept clean and running. They have usually been cut and laid for the purpose of draining the land, and if they are not maintained, do not serve the purpose. Drain outfalls should be kept open at all times. Ditches should be protected from livestock treading them in, with barbed wire fencing.

17.6 Farming standards

Farming standards have much effect on farm values. Clearly, if a farm is farmed to a high standard and is highly productive, neat and tidy, it should be worth more than a comparable holding which has been neglected and needs considerable expenditure and effort to bring back into full and proper production. Often, the fixed equipment is neglected and it has no modern buildings. Again, the land often has been allowed to become weed-infested, impoverished and lacking in required levels of potash, phosphate and lime to sustain proper plant growth. Hedges, fences, ditches, gates, culverts, drains and roads may be neglected. The cost of improving all or any of these to proper standards is substantial and often it may take three to five years, or even more, to complete the work.

17.7 Services

The availability of all main services is a great asset. However, some of these main services can be costly in use. If a dairy farm is entirely dependent on mains water supplies and a large herd is kept, water charges will be considerable. Private supplies, which cost very little or nothing, avoid considerable expenditure. Likewise, the availability of irrigation facilities from streams and lakes has considerable value and can increase productivity dramatically despite the charges water authorities make for such. Effluent charges, where effluent is discharged into main sewers on dairy, pig and specialist farms on the outskirts of towns, can also be very high. If effluent can be treated on the farm, as is often the case, this will again avoid considerable expense.

17.8 Sporting

The existence of good sporting rights on a farm, in particular well placed woods, coppices and dingles that provide cover for game and can create a good shoot, is of much value. If some of these woods are so placed that they can provide high sporting birds, this will again improve the shoot. Likewise, the presence of duck flight ponds or small lakes that can be so developed is a great attraction. Even if a farm has all these facilities and no game, a good shoot can be created by rearing game and providing further cover by growing root crops, such as kale, swedes, and possibly leaving small areas of unharvested grain, and also planting small corners up.

If a holding has a trout stream or a lake which can be stocked or, better still, some river fishing (with good pools for holding the fish) this, again, is an attraction and an asset.

Sometimes, farms are sold with the shooting and other sporting rights reserved. This is a disadvantage, apart from the interference caused by persons exercising these rights.

17.9 Development value and cottages

There is sometimes a possibility that a farm may have some development potential if it has land adjacent to a town or village. Provided this is a distinct possibility and not a far distant hope, it will add value to a farm. Likewise, the presence of a number of cottages (provided they are not restricted in use to agricultural occupation) can be an asset, particularly as often these can be sold off at good prices even if derelict.

In recent years considerable value has been added to many farms which have barns and other redundant buildings that can be converted to dwellings and holiday lets.

17.10 Woodland

An area of woodland can be most useful on a farm quite apart from providing cover for game, shade and shelter and for use as a windbreak. Farmers are generally in need of timber for fencing posts, gate posts, etc. Material of this kind is expensive to buy and a good and suitable free farm supply is thus an asset, provided the area concerned is not disproportionate to the size of the holding. Since the use, in the last two decades of woodburning stoves, an adequate free supply of firewood is of great value.

Generally, the most valuable timber for a farmer is hardwood such as oak and chestnut (for fencing posts), and the best softwoods (for rails and conversion), larch and spruce.

It is usually prudent for a farmer to plant unproductive areas, such as dingles and odd corners with quick-growing softwoods and wet areas with poplar. Apart from the timber value the shade and shelter provided, this will also be a conservation measure.

17.11 Easements, wayleaves, rights of way, etc

Rights of way, easements including pipeline easements, electricity board lines, restrictions, etc can have a depreciating effect on capital and rental values and their effect should always be considered when valuing a freehold or assessing its rental value.

17.12 General

A few matters considered as a plus to be of value on a farm are set out below.

17.12.1 A good hard internal spine road, or road to give easy access to fields, and if in the form of a lane, this is ideal for driving stock along without damaging crops.

17.12.2 Natural water supplies, such as intersecting or bordering streams. Adequate rainfall.

17.12.3 Good shade and shelter, which is essential on livestock farms.

17.12.4 That the farm has a good layout with a centrally placed homestead with not too much of a pull up towards the homestead. Steep roadways of this kind can be difficult at almost any time, but particularly in the winter and, of course, are energy-consuming.

17.12.5 That the homestead should, if at all possible be sheltered and protected from the prevailing wind. If it is very exposed, it will be a cold place in which to live.

17.12.6 The land not being too exposed to prevailing winds, eg a farm on a high altitude with no protecting hills or windbreaks.

17.12.7 The possibility of splitting a farm so that a portion can be sold off, if desired, may well enable a potential purchaser to conclude a deal.

17.12.8 The quality of the soil always, in the final analysis, determines the value of a farm. Versatile soils are favoured, being not too heavy and not too light, but probably a deep loam with some body in it, ie tending to be slightly more heavy than light. Good land can be found everywhere, but some of the most fertile and popular agriculturally favoured areas of the country are Herefordshire, Shropshire, Taunton Vale, Cheshire, The Fens, Vale of York, Fife, Angus and Ormskirk area of Lancashire. The high prices realised for farms in these districts indicates their popularity.

The Agricultural Land Classification map should always be studied. To date, these maps are provisional editions only and may yet be amended. The land is placed in five categories with Grade I and II the best, and with only minor limitations. Grade III is also good with only moderate limitations. Grade IV has severe limitations and Grade V is usually rough grazing. Generally speaking, these maps are reasonably accurate but they are on a small scale and deal with fairly large blocks. Often, some really good small areas of land are not indicated as they should be, and the converse also occurs.

17.12.9 It is now obvious that holdings possessing quotas, whether they are sugar beet or milk quotas, will in the future have considerable attraction to a farmer who is interested in producing a commodity for which the holding has a quota. This is best illustrated by stating that prior to the introduction of milk quotas, there was no restriction on milk being produced and sold off any farm provided all regulations were complied with. This is no longer possible. A quota can usually be purchased but is expensive. The amount of the quota is also very relevant — clearly a holding with a large sugar beet or milk quota will be more attractive and worth a great deal to any buyer than similar holdings with small or no quotas. This has had far-reaching effects on capital values and rental values.

17.12.10 The possibility of diversification, eg running a farm shop, caravan site, bed and breakfast business, etc.

17.12.11 Single farm payment

If the land is registered with DEFRA and includes single farm payment it normally has a much increased value, especially if the payment is fairly substantial.

Inventories (Records)

18.1 The accurate preparation of inventories is important on the completion of any agricultural valuation. A good inventory is of particular importance to a tenant farmer when he enters a farm, and this can be very useful when he vacates. The inventory will prove what he paid for, eg tenant's fixtures, tenant's pastures, improvements and also the cropping of some land on entry. The CAAV recommend that inventories should always be prepared by the incomer's valuer as he is more likely to be concerned about its accuracy than the outgoer's valuer.

18.2 While giving too much detail may possibly result in questions being asked and sometimes controversy between a client and his valuer (this should not arise provided there is proper consultation between them), it is nevertheless important to set matters out in a fairly detailed manner, particularly:

(i)   values given in respect of any equipment taken over. If this is not done, this information will have to be divulged at a later date for accountancy purposes. It is not necessary to give individual prices, except perhaps for tractors and major items of equipment

(ii) details of tenant's pastures, ie OS numbers of areas, when laid down, and duration of the ley

(iii) tenant's fixtures and improvements should be clearly identified

(iv) where dilapidations are valued and offset, it is advisable to give as an annex to the inventory, details of what items have been allowed for and with the total compensation allowed, but not individual prices. This will show to the incomer what has been allowed for and what he is expected to remedy.

18.3 It is not particularly important to state what the previous crops grown prior to those actually valued were. The incomer can ascertain this from the outgoer himself, and rotations are now seldom practised.

18.4 The inventory should give the effective date of the valuation, the date when the actual valuation was carried out, and should be signed by both valuers. The incoming tenant should keep the inventory in a safe place, preferably with the tenancy agreement, as it could well be needed one day.

Specimen inventory

INVENTORY AND VALUATION
of
IMPROVEMENTS, TENANT RIGHT
FIXTURES AND EQUIPMENT
at
LOWER MOOR FARM, PENTLAND
in the County of GWENT

From: A.H. MOOR ESQ   Outgoing Tenant
To: J. H. LEWIS   Ingoing Tenant

As at 2 February 2008
Valuations taken and made 31 January 2008

1.   PRODUCE
1.1 **Hay** — at consuming value
Five bays baled 2007 Seeds Hay in Dutch Barn
Part bay baled 2006 Meadow Hay in Wergins Barn

1.2 **Straw** — at consuming value
Two bays baled 2007 Winter Barley
Straw in Dutch Barn
28 big bales 2007 Winter Wheat Straw in OS No 123

1.3 **Silage** — valued in accordance with the basis given in NP 168, CAAV (as amended) Part used clamp, 2007 grass silage part used clamp 2007 Maize silage

1.4 **Roots** — Clamp Golden Tankard Mangolds

2. **GROWING CROPS**
(All at cost of cultivations, seeds and, where applicable, sprays and enhancement value)

2.1 **Stubble Turnips** — OS No 345. 3.56 ha
The growing crop of stubble turnips on 2ha

2.2 **Swedes** — OS No 365. 4.81 ha
The growing crop of swedes on 2ha

2.3 **Winter Wheat** — OS No 567.11.71ha
The growing crop of Avalon Winter Wheat

2.4 **Winter Barley** — OS No 789.13.45ha
The growing crop of Igri Winter Barley

3. **CULTIVATIONS**
3.1 **Ploughing** — Ploughing in OS No 891.10.2ha
Ploughing in OS No 894.11.21ha

4. **LABOUR TO FARM YARD MANURE**
Heap carted to OS No 788

5. **TENANT'S PASTURES**
5.1 OS No 346. 7.81ha — 4-5 year ley undersown Barley 2005
5.2 OS No 347.10.04ha — 3-year ley direct seeded autumn 2007
5.3 OS No 349. 8.56ha — Permanent ley direct seeded autumn 2006
5.4 OS No 350. 7.20ha — Permanent ley sown 2000

6. **SOD VALUES** on OS Nos 412 and 418 — 10.8ha

7. **UNEXHAUSTED MANURIAL VALUES**
UNEXHAUSTED VALUE OF LIME APPLIED
RESIDUAL VALUE OF FEEDING STUFFS CONSUMED
The unexhausted value of fertiliser and lime applied to the land and the residual value of purchased feeding stuffs and home-grown corn consumed on the holding

8. **TENANT'S IMPROVEMENTS**
8.1 Four bay steel and asbestos Dutch barn 90' × 30' erected in 1993
8.2 Tile drains laid in OS No 350 (serving 2.5ha) carried out in 2005

9. **TENANT'S FIXTURES** (taken over and paid for by the Incoming Tenant)
9.1 Sheep dip, foot bath, four pens with 7-rail tanalised post and rail fencing, together with all concreting, water service and drains. Constructed in 1994.

9.2 Post and pig netting fencing with two strands of barbed wire on tanalised post and rail fencing — 410 m run dividing OS No 886. Erected in 1996.

SUMMARY

1.	Produce	£12,078.50
2.	Growing crops	4,621.40
3.	Cultivations	450.00
4.	Labour to farmyard manure	271.20
5.	Tenant's pastures	2,420.00
6.	Sod values	387.00
7.	Unexhausted manurial values, Unexhausted lime and residual value of feeding stuffs consumed	1,281.25
8	Tenant's improvements	2,480.00
9	Tenant's fixtures	870.00
		£24,859.35
	Less an allowance for DILAPIDATIONS as shown on attached ANNEX*	9,241.00
	Net Valuation	£15,618.35

Net Valuation
Dated: this 20th day of February 2008

We, the undersigned, having inspected the above, DO VALUE, the same in the sums given.

. . . . FRICS, FAAV
Chartered Surveyor,
Acting on behalf of the Outgoing Tenant

. . . . FRICS, FAAV
Chartered Surveyor
Acting on behalf of the Incoming Tenant

*NOTE A list of the dilapidations is normally attached to the Inventory. This shows the items of claim excluding the figures agreed per item — merely a total is given at the end.

This list is not attached to this particular example inventory, but would more or less take the form of the items of claim (excepting any items deleted on negotiations, given in the example schedule of Dilapidations in Chapter 11). A total of the agreed claim is given.

Compulsory Purchase Claims

19.1 The past 50 years have seen a dramatic increase in claims arising on compulsory purchase, largely on land being acquired for housing, schools, sewerage works, motorways, roads, road improvements etc. The law is governed by the following acts.

19.1.1 Land Compensation Act 1961

Section 5 lays down six rules for assessing compensation as follows:

Rule 1	No allowance is to be made because the acquisition is compulsory
Rule 2	The value is to be the open market value assuming a willing seller
Rule 3	Special suitability for a statutory purpose or where there is no market apart from the special needs of a particular purchaser is to be disregarded
Rule 4	Increase in value for uses contrary to the law is to be disregarded
Rule 5	Special and rare cases only can be dealt with on a cost of reinstatement basis (generally where it is used for a purpose for which there is no market eg churches)
Rule 6	Rule 2 above not to affect assessment of compensation for disturbance on other matter not based on the value of the land.

19.1.2 Compulsory Purchase Act 1965

Compensation for injurious affection and severance — section 7. Compensation for yearly tenants or tenants for less than a year — section 20 (see also the 1973 Act).

19.1.3 Land Compensation Act 1973

Claims for road noise and other physical factors (eg smell, lighting etc.) where no land is taken from the claimant — Part I.

Severance of an agricultural holding — sections 53–59 and 61.

Expenses of removal, home loss payment — rehousing etc.

Tenant's security of tenure — section 48.

Planning blight of an agricultural unit, now in sections 158–160 of the Town and Country Planning Act 1990.

19.1.4 Agriculture (Miscellaneous Provisions) Act 1968 section 12

Entitlement to re-organisation payment — see section 48(5) of the 1973 Act. Compensation of four times the rent (or apportionment). This additional amount is tax free, but has to be deducted from the value of the tenancy interest (see 19.3(a)). Valuers sometimes fail to appreciate that it is the rent actually payable and not rental value.

19.1.5 Planning and Compensation Act 1991

Additional home loss compensation for residential owner-occupiers — sections 68 and 69.

Notices to treat ceasing to have effect after three years unless compensation agreed, possession taken etc — section 67.

Revision of certificates of appropriate alternative development — section 65.

Planning assumptions in connection with highway schemes — section 64.

Availability of interest payments — section 80 and schedule 18.

19.1.6 The Planning and Compulsory Purchase Act 2004

This Act introduced two new types of compulsory purchase payments — basic loss payment (max £75,000) and occupiers loss payment (max £25,000). Both were inserted by this Act as section 33A and 33B of the LCA 1973. Farm loss payments under section 34 of the LCA 1973 have been abolished and replaced by these new payments.

19.2 Owner/occupier claims

Owner/occupier claims broadly cover the points set out below.

19.2.1 Open market value

A claimant is entitled to the open market of the land acquired at the time of the claim being settled or the date of possession in stake, whichever is earlier. They are also entitled in assessing the value of the acquired holding to consider splitting the property in a prudent fashion and valuing in units, if this produces a higher total market value. Such an approach may, however, affect the claim for severance or injurious affection. The possibility of development value should be considered and where appropriate an application for a certificate of alternative development made. Where granted, this can add considerably to the claim. (Section 17 of the 1961 Act, as amended by the Local Government Planning and Land Act 1980 and the Planning and Compensation Act 1991.) In certain circumstances, the acquiring authority may be able to make a set-off for betterment to other land of the claimants, arising directly from the scheme, eg where the LPA would grant permission for a petrol filling station, but only in consequence of the road improvement scheme.

The treatment of milk quota can be a very important matter where part of a dairy farm is acquired. In the vast majority of cases, the claimants (owner, occupier, landlord or tenant) will agree that the entirety of the quota will remain on the holding. If this was not the case an apportionment is made in accordance with the *Pucknowle Farms Ltd* v *Kane* [1985] 2 EGLR 8 decision.

19.2.2 Severance and injurious affection

In addition, claims for severance and injurious affection may well arise, eg a farm split by road and injury to the value of the farm as a unit or parts of it including proximity of a new road to the dwelling house, interference with sporting rights, depreciation arising on loss of land resultant in certain fixed equipment being superfluous to the needs of that holding, eg 30 acres of a 170-acre farm acquired. A specialist milking set-up can no longer be used to optimum capacity. Also, additional costs of cultivations, spraying, harvesting, etc, which arise where parts of fields are acquired. Depreciation in the value of the claimant's retained land by virtue of added fencing liability comes under this heading (see *Cuthbert* v *Secretary of State for the Environment* [1979] 2 EGLR 183).

Care needs to be taken to ensure that losses for severance and injurious affection are assessed on the basis of the depreciation in the market value of the claimant's retained interest. Arguably the market will take little account of a farm with over-capacity in the milking parlour but these losses will be reflected in a before and after approach. The Lands Tribunal have rejected claims on a number of occasions where these have been based upon capitalised future costs.

19.2.3 Disturbance

This generally forms a major part of a road acquisition and some other similar cases. However, these losses must be a direct result of the acquisition. Negligence by the contractor should normally lead to a claim against him and not the acquiring authority: see, however, 19.5 below. Possible items of claims include:

(i)   loss of crops, cultivations, pasture, UMVs, RVFS, RV Lime
(ii)   loss of profits on a crop almost ready for harvesting (intrinsic profitability is reflected in the value of the land)
(iii)   expenses incurred in moving or moving stock
(iv)   damage to crops, eg dust settled on a silage crop
(v)   damage resultant in rock being blasted onto adjacent fields
(vi)   cost of rounding up stock which has strayed because the fencing provided was not stockproof
(vii)   injury caused to stock as a result of the works, eg wire left about
(viii) where herbage seeds are grown, the loss arising where certain fields or part fields, alongside the new road, can no longer be

certified (and thus used for such production) owing to pollination by grasses growing on the new road verges

(ix) cost of extra cleaning of dwellings including carpets and curtains arising from dust emanating from the works in progress

(x) flood damage arising as a result of the works

(xi) a neighbour's bull gets into a field (owing to defective road fencing) containing young beef heifers and serves them

(xii) damage to drains

(xiii) interference with a shoot on a temporary basis (note a claim for permanent damage will arise under injurious affection)

(xiv) loss on forced sale of stock

(xv) claimant's time properly spent with agents, contractors, engineers, etc, or indeed any time spent by the claimant mitigating his loss

(xvi) claims for redundancy payments arising where employees are made redundant consequent on the acquisition

(xvii) cost of re-painting and decoration occasioned by dust arising from the works

(xviii) loss where the works interfered with a farm shop or similar retail business

(ixi) expenses properly incurred including legal and surveyors' fees in searching for and purchasing an alternative property (where the whole is acquired)

(xx) severance of water supplies

(xxi) blasting causing broken windows and fractures to walls and buildings.

All the above matters have been encountered by valuers acting on behalf of claimants, particularly in the case of acquisitions for new roads. It is impossible to foresee the extent of the claim for disturbance when completing the claim in reply to a notice to treat but it is advisable that a reservation should be made that this additional claim will follow and will be quantified when the work is complete. The golden rule is to take a record of condition prior to entry, preferably agreed with the authority, together with good photographs and also to advise the claimant to keep a detailed diary of every incident arising and the time he has spent on the matter. The total hours used are added up at the end of the work and a suitable claim made for the loss.

19.2.4 Basic loss and occupiers loss payments

Farm loss payments are now abolished and replaced by basic loss and occupiers loss payments (Planning and Compulsory Purchase Act 2004). Basic loss payment is 7.5% of the value of the interest acquired — maximum £75,000. The maximum occupiers loss payment is £25,000.

19.2.5 Home loss payment

This arises when a home is lost by compulsory acquisition, after being occupied by a claimant for at least one year. The amount of the payment is (under section 30 of the 1973 Act as amended) 10% of the market value of the claimant's interest in the dwelling subject to a maximum of £15,000 and a minimum of £1,500. It is expected that now domestic rates have been abolished, a flat rate will be fixed. The claim has to be lodged within six months of dispossession.

19.2.6 Accommodation works

These are works carried out on the claimant's remaining land in partial mitigation of the claim for injurious affection. In strictness there is no right to accommodation works but it is normal to negotiate accommodation works as part of the compensation and these may include the following.

(a) Replacement of fences, replanting of hedges and maintaining these until established and protecting the same. The loss in market value arising from future burden of maintenance should be reflected in the injurious affection of the land remaining. The Department of Transport will accept liability for maintenance of motorway fences but no others. It is therefore important, as fences perish and hedges need maintenance, protecting and laying, fully to reflect the loss and future liability. It has been held that the loss falls under Rule 2 (market value loss) and not disturbance as mentioned in Rule 6 and this item of claim should be included under injurious affection (*Cuthbert* v *Secretary of State for the Environment* [1979] 2 EGLR 183).

(b) Provision of gates. It is considered new gates should always be at least 4m wide, galvanised and the gateways stoned.

(c) Removal of hedgerows where justified where odd areas of fields are left and these can be let into adjoining fields but this will mitigate the claim for severance. Planning consent is needed for this work.

(d) Provision of water supplies, troughs and meters (and pay the proper loss resulting from any additional meters installed in consequence of the acquisition) where water supplies are severed. Where supply pipes have to pass under a new road, it is wise to have these in ducts or sleeves so that such pipes can be withdrawn and replaced if ever considered necessary. It is important to note that these ducts or sleeves should be of ample size to withdraw and insert pipes. Cases have arisen in the early days of motorway construction of such ducts etc, being too small, the new road has since settled, the ducts severed and it is now impossible to replace any services in need of replacement.

(e) Continuance of severed land drains, including provision, where appropriate, of a header drain to discharge into the roadside drains (full rights for such discharge may be reserved in the conveyance).

Replanting of shelter belts, specimen trees, ornamental trees, etc, on adjacent land retained. This will also cover maintenance until established. In practice, the acquiring authority may pay the claimant to do the work himself. (It may also be possible to arrange for works to be done on the land acquired, but this is by agreement and not, strictly speaking, accommodation works.)

(g) Noise insulation and double glazing, special ventilation and, if necessary, Venetian blinds.

(h) Provision of access to severed land (partly by means of land acquired from neighbours) by the provision of over-bridges or under-passes. Note that in the case of the latter, it is important that they should have ample headroom to take high loads and allow tractors with high cabs to pass and the former should be of adequate width. Also, many under-passes are sunk into the ground and problems have arisen with these getting flooded in wet weather. Such a bridge or pass will also involve consideration of re-sorting existing farm roads to co-ordinate with the crossing point. The main structure of the overbridge or underpass is not an accommodation as it is constructed on acquired land and the claimant will need to reserve an easement of way with liability for future maintenance passing to the acquiring authority. Access aprons and ramps, however, are constructed on the claimant's land and properly constitute accommodation works.

(i)   Making good damage to part structure acquired, eg part buildings taken.

Note: The claimant does not have an entitlement to a like for like replacement and if agreement cannot be reached the acquiring authority will pay compensation rather than carrying out the accommodation works.

19.2.7 Fees

Ryde's Scale is now abolished. Surveyors fees should now generally relate to the reasonable cost of undertaking the work and reflect time, expertise and effort required. Travelling and out of pocket expenses should be claimed and VAT. Fees should also be claimed on the value of accommodation works.

It is very important that the valuer should keep a detailed record of his visits and expenses. Also proper legal costs. Legal costs are payable by statue and need not be the subject of claim but it is as well to mention them.

19.2.8 Interest and advance payments

This is payable at the statutory rate on the unpaid balance of the settled claim. Advance payments of compensation (90% of the acquiring authority's estimate and payable when entry is made) may be available under section 52 of the 1973 Act. Continuing applications for further advances should be made if the acquiring authority increase their estimate of the proper compensation. Under section 52A the claimant is entitled to a payment of recovered interest at the same time as any advance payment is made.

19.2.9 Plan

It may be a great nuisance to a landowner to have, say, 4,500m^2 of land acquired from, say, three fields. He does not know the individual area of the remainder. The acquiring authority should therefore be requested to prepare and supply the claimant with a plan of the area of the farm adjacent to the acquisition, where part OS numbers have been acquired. This plan should show the individual area remaining in acres and hectares of each affected enclosure.

There is no obligation on the acquiring authority to provide plans but in suitable cases the costs in preparing such plans would form a claim under Rule 6.

19.3 Tenant's claims

Tenant's claims are governed by the same Acts as previously mentioned and reference is made, in particular, to the decision given in the cases of *Wakerley* v *St Edmundsbury Borough Council* [1979] 1 EGLR 19 and *Anderson* v *Moray District Council* 1978 STL 37. The matter is complex but very simply the tenant is entitled under section 20 of the 1965 Act to:

(a) the value of the unexpired term of the tenant's interest (in his tenancy and ignoring any prospect of a notice to quit founded on the authority's scheme)
(b) the payment that would be made to him by an incoming tenant, eg value of growing crops, UMVs, RVF stuffs, unexpended value of lime applied, tenant's pastures, tenant's improvements, etc
(c) loss or injury sustained, such as disturbance, loss on forced sale of stock, removal expenses, home loss payment (if living in the house continuously for one year — see the 1973 Act), basic loss payment (max £75,000), or occupiers loss payment (max £25,000)
(d) severance where this can be proved and injurious affection to the remainder of the farm.

Any contractual bar on assignment is to be ignored because this is overridden by the concept of a notional sale, Rule 2 (19.1.1).

In order to avoid duplication of claims, section 48(5) of the 1973 Act provides that the above compensation is subject to a deduction of the payment of four times the annual rent paid (or the apportionment) in respect of the land taken (section 12 of the 1968 Act). [For basis of apportionment see 19.6.4.] This of course applies where the claim for the interest is in excess of four times the rent.

Care should be taken to note whether the dispossession is by notice to quit (under the power of resumption of possession in the tenancy agreement, if it exists), ie where the authority has already bought the freehold, or by notice of entry (see section 11 of the 1965 Act).

Where entry is made by notice to quit the tenant is not entitled to be paid for the value of his interest (a) above [but see below] being merely entitled under the Agricultural Holdings Act 1986, section 60 to a minimum of one year's rent and a maximum of two years but he will, in addition, get the four years' rent re-organisation payment under section 12 of the 1968 Act or the occupiers loss payment. However, if it is to his advantage he may claim to go out under his tenancy agreement in which case a notice must be served under section 59(2) of the 1973 Act. In this event the claim is made under section 20, as above, but should not then include an item for section 60 of the Agricultural Holdings Act 1986.

Where entry is made by notice of entry, the claim is assessed under section 20 of the Compulsory Purchase Act 1965 and comprises all of items (a), (b), (c) and (d) above, subject to the deduction under section 48(5) of the 1973 Act. The section 12 (1968 Act) payment of four times the rent is thus preserved as a tax-free payment but the deduction is made to prevent double counting.

19.4 Valuation of the tenant's unexpired interest

This is a very difficult valuation to estimate especially in the case of a strip of land taken from a farm. If a whole farm is acquired it will prove easier to assess as evidence can be found of compensation paid by landlords to their tenants for vacating and there are cases known where the compensation has ranged from merely offsetting dilapidations against tenant right to well over £1,500 per acre being paid for possession being given. The following factors need to be considered.

(a) The age of the tenant. If he is a young man say in his thirties, and has no desire to vacate, he is most unlikely to consider vacating unless very substantial compensation is paid. His chances of obtaining another rented farm are very remote and he will probably lose his living. If, on the other hand, the tenant is near or over retiring age and has no family hoping to succeed to the tenancy, the value of the interest is substantially diminished, especially if his health is not good (but see (f) below). This is particularly important where the method of valuation is founded on the surrender value to the landlord.

Most acquiring authorities would argue that it is necessary to assess market value in accordance with Rule 2 and hence the necessity to envisage a willing seller. Since the 1977 Act (now case G86A) a landlord has been unable to serve an uncontestable notice to quit on the death of the original tenant, where the tenancy has been assigned. In these circumstances the age of the tenant has assumed less importance.

(b) The possibility of a member of the tenant's family being able to apply for tenancy succession. Although such a family member has no legal interest whatever in any claim, this is bound to be a factor that would influence the attitude of the tenant and it is contended that it has a definite bearing on the value of the tenant's interest.

(c) How important any given area of land is to a particular tenant. If only a smallholder with the farming being of a high intensive standard, any area of land acquired would be far more important to him than in the case of the tenant of a large holding which is not intensively farmed. In other words, the latter might be more likely to accept a relatively smaller compensation for vacating a given area of land than the former.

(d) Whether there is any possibility (other than in the case of the acquisition) of the landlord being able to obtain planning consent for any other user than agriculture and consequently obtain possession of the area of land involved. [Note: The Act requires the valuer to leave out of account the prospect of a notice to quit founded on the actual scheme underlying the acquisition — section 48 1973 Act.]

(e) Is the tenant enjoying a profit rental? If on a long lease with no rental revision provisions he may be, but in other cases, the maximum period of profit rental he could claim would be three years.

The extent of the tenant's improvements and the value to be disregarded under schedule 22(l)(a) of the 1986 Act is relevant.

(f) The possibility of assignment (even if prevented by the tenancy agreement). It is necessary to consider a hypothetical assignment or surrender to the landlord (ie the law presumes a willing seller, which in turn means leaving out of account any bar on assignment).

To summarise, the tenant is compensated for his interest as if it existed in a no-scheme world, being at risk of dispossession solely in so far as his tenancy agreement and the statutes would allow, ie the valuer must ignore any possibility of a notice to quit founded on the authority's scheme.

Several different approaches have been taken by valuers to assess the value of the tenant's interest. A Lands Tribunal case of significance is the *Wakerley* case (*Wakerley* v *St Edmundsbury Borough Council* [1977] 1 EGLR 158 which went to the Court of Appeal [1979] 1 EGLR 19). This, however is not a particularly helpful case and may not have another parallel. The case of *Anderson* v *Moray District Council* (1980) decided that since the land in question had development potential, an effective notice to quit could be served on the tenant and this had the effect of reducing his security of tenure. Among the suggested approaches made by valuers are:

(a) a proportion of the difference between the vacant possession value and the value of the reversion
(b) a proportion of the vacant possession value of the area acquired
(c) a proportion of the difference between the vacant possession value of the freehold interest and its value subject to the life tenancy. This will have regard to the age of the tenant
(d) capitalising the profit rent for the remainder of the tenant's lifetime
(e) capitalising profits for a reasonable period at a rate which reflects risk taking
(f) basic loss payment (7.5% of the value of the interest acquired up to a maximum payment of £75,000 (new section 33A of the Land Compensation Act 1973)
(g) occupier's loss payment (maximum of £25,000 section 107 of the Planning and Compulsory Purchase Act 2004.

It has been stated that valuation is not an exact science. This statement is nowhere more true than when valuing an agricultural tenant's unexpired interest. In many cases there will be considerable variance between the claimant and his valuer's views and those of the district valuer and often the claimant will be dissatisfied with the final compensation agreed or awarded. Until tested in the market, there is never a definite value, but rather a range of values between which the answer may reasonably be expected to be.

The CAAV discussion paper entitled *Valuation of Agricultural Tenancies for Taxation Purposes* dated June 1997 refers to many considerations of relevance to this valuation issue.

19.5 Procedural points

From experience it is desirable for a record of condition of the land to be prepared and agreed in road acquisitions, prior to entry. Photographs should be taken. These can be very useful at a later date.

It is considered that to start with, no claims should be made against contractors. In the first instance the matter should be referred to the resident engineer but all claims should initially be lodged against the acquiring authority who are ultimately responsible.

Contractors frequently wish to hire sites for soil dumps, spoil removed, working space, depots, etc. They are usually prepared to pay what might appear to be a good consideration for such uses. These sites should always be licensed and not let, a proper agreement entered into and full payment made prior to entry. Contractors should be responsible for fencing and full reinstatement. When using land for depots, the long-term damage caused can be considerable since contractors usually lay hardcore, erect hut bases, lay concrete and form roads. It should be stipulated that the top soil should be removed and stored prior to any work being done. All debris should be removed on completion, including all hardcore, stones, rock, etc, the land subsoiled at least three times, top soil replaced, fertilised and re-seeded. The consideration charged should reflect that the land affected is unlikely to be fully productive for several years. Hence the inference above 'might appear to be good consideration' is not necessarily so. The acquiring authority is not concerned with these private arrangements.

Example claims

Owner/occupiers claim

Moorend Court is an 85-ha (210-acre) farm about 0.75 mile from a market town in the south-west. It has a period, five-bedroomed Georgian residence and extensive, principally modern buildings (many of which have been erected in the last 10 years). It also has two semi-detached farm workers' cottages. 110 pedigree Friesian cows are kept, together with about 40 followers. There is also a bull beef unit producing about 50–60 bulls a year. About 20ha (50 acres) of cereals are grown. The land is all Grade 2 and the soil is free-draining loam derived from old red sandstone. The farm has a primary wholesale milk quota of 556,000 litres.

The Highway Agency are about to construct a single carriageway by-pass which affects Moorend Court as shown on the plan. All legal formalities have been complied with and the compulsory purchase order in respect of the scheme was confirmed on 21 December 2004. Entry was taken on 1 April 2005. The by-pass will severely affect Moorend Court, severing a large part of the land from the homestead. The new road will be within 110m of the residence.

At various meetings held between the county engineer (acting on behalf of the Department of Transport), the resident engineer, the district valuer, the claimant and his agent, it has been agreed:

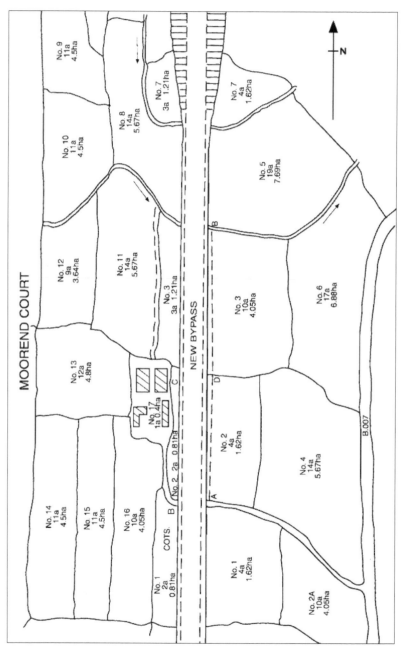

(a) an access bridge will be built across the by-pass to give access to the homestead from road B.007

(b) a concrete farm road will be built from points A to B

(c) the existing water main supply will be re-laid in a 300mm diameter sleeve under the by-pass between points C and D

(d) a noise bund will be constructed on the new boundary opposite the residence

(e) all drains affected will be connected to the surface water drains to be constructed alongside the new carriageway

(f) a two-row staggered quickthorn hedge is to be planted on the new boundary and is to be protected on the field sides with a pig netting fence on tanalised posts with two strands of barbed wire

(g) the milk quota will be retained by the claimant (*Puncknowle* case).

The area of land acquired is 6.07ha (15 acres). 33ha (82 acres) of the farm will be severed from the homestead. Licences are also being taken to re-align the brook in OS No 7 and bordering OS Nos 5 and 8.

The duration of the contract was to be one year from the date of entry. However, for a variety of reasons, the contractors were on the land several months over the completion date. The over-bridge took much longer to complete than anticipated and there were problems of access to the severed land with resultant delays in fertiliser application, silage making and inadequate access to get the dairy herd across to the set stocked grazing pastures. The temporary fencing erected was not stockproof and was frequently damaged by the contractors and stock got out. Further damage was caused when entry was taken to remove this and plant the quickthorn hedges. Damage was also caused when the contractors entered to locate a blocked drain which caused flooding. There were major problems with dust resultant from earth moving in 2006. This dust affected the residence, gardens, all buildings, pastures and silage cuts. The spring of 2007 was exceedingly wet and very muddy conditions existed until 1 June. The water supply to both the homestead and the severed land was cut and milk yields lost. Rock blasting resulted in a bill of £1,210 for damages to a precision chop silage harvester and silage making was delayed for two days. 1ha (1.5 acres) of grazing was lost owing to inadequate fencing for 10 months.

STATEMENT OF CLAIM

by

Mr JR Yeoman (Freeholder and Occupier)

against

The Department of Transport

following the acquisition of part of

MOOREND COURT, WESTINGFIFLD, ANY COUNTY

for the construction of the Westingfield By-Pass

March 2005   Prepared by: WILLIAM DAVIES & REES
                            Agricultural Valuers & Chartered Surveyors
                            Dolanog
                            Welshpool, Powys

Pt OS Nos 1, 2, 3,5 & 7 — Area 6.07ha (15 acres)

1.   LAND TAKEN

1.1 Freehold Interest Acquired

The Acquiring Authority to pay for the freehold
interest acquired — 5.07ha @ £12,500   £63,375

1.2 Licence

The Acquiring Authority to pay for the user of
the area of 0.6ha used as a licence for the
duration of the contract ie 2 growing seasons.
The sum of   £2,000

Total for land taken   £65,375

2.   SEVERANCE

Area severed east of the by-pass is a total of 33ha
(82 acres). Cow grazing pastures have previously
been OS Nos 2, 3, 11, 5, 7 and 8. It will no longer
be practicable to use OS 5 and 7 (east) and
OS 9 will now have to be used instead.
There will also be additional travelling time for
men and equipment to collect the cattle from
OS No 9 and for husbandry in the severed land.
Cattle will also be travelling twice a day for part
of the grazing season into OS Nos 5 and 7 (east)
and will have to be cleaned up at wet times apart
from the damaging effect of cow dung on
tarmacadam and also general wear and tear.
Additional estimated time taken due to the
by-pass severance is 150 hours per annum.
The Acquiring Authority to pay loss in value of
33ha due to severance at £1,000ha.

Total severance   £33,000

3.   INJURIOUS AFFECTION

3.1 The residence is severely affected by traffic
noise, fumes, car lights shining at night, dust
and its view is severely impaired. Current value
of residence as part of the whole farm is
estimated at £550,000. Loss 30%   £165,000

3.2 The pair of cottages are similarly affected.
Current value of each is £120,000. Loss 30% £72,000

3.3 The efficient running of the holding is also
severely affected by the proximity of the
by-pass. Certain areas cannot, in future,
be used for herbage seed production and
in all approximately 12ha will be affected by
noise, fumes, dust etc. There is also the risk
of stock getting out onto the by-pass.
The occupation of the farm buildings are
also effected.

Loss:

3.3.1 12 ha capital value of £12,500
per ha, depreciating by 10% £15,000

3.3.2 Buildings with capital value of
£400,000 depreciating by 10% £40,000

3.3.3 Buildings and equipment with a capital
value of £108,000 (£100,000 buildings;
£8,000 equipment) will become
redundant. Loss £10,800

3.3.4 1020m of fencing has to be maintained.
This will need annual maintenance in
perpetuity and the hedge, when of
sufficient height (with about 10 years'
growth) will need laying.
Depreciation in value in the remaining
79ha (195 acres) remaining in the
claimant's ownership @ £750 per ha £59,250

Total Injurious Affection £362,050

4. **TIMBER & AMENITY**
Loss of 10 mature oak trees and 12
specimen copper beech trees in
OS 3 (west)

4.1 30 cu m oak and beech @ £100 £3,000

4.2 Loss of amenity and shelter 12 trees @ £100 £1,200

Total claim for Timber £4,200

Total Amount of Claim **£464,625**

5. **DISTURBANCE AND CROP LOSSES ETC**
   This is to be the subject of an additional claim to be made when the contract is completed.

6. **ACCOMMODATION WORKS**
   The Acquiring Authority to execute at their expense and to the claimant's complete satisfaction, the following accommodation works.

   6.1 Plant a staggered two-row quickthorn hedge on the new boundary protected on the field side by pig netting and two strands of barbed wire on tanalised posts and maintain, including annual weeding, until fully established.

   6.2 Grub out and level the hedgerow between OS Nos 7 (west) and 8; OS 4 and 2; 11 and 3 (west) and 7 and 5 (east).

   6.3 Remove the redundant farm road in OS No 11 and fill the roadway with top soil.

   6.4 Connect all exposed drains to the new roadside drains to be laid. Also to repair all drains affected by the works and to allow, at all times, any new drains laid by the claimant (or his successors in title) on the holding to discharge into the roadside drains.

   6.5 Supply, fix and connect to the main water supply water troughs to serve OS Nos 3 and 5. This will entail laying a main from the existing farm supply and providing a PVC sleeve to take the main under the road — 300mm diameter encased in 150mm concrete.

   6.6 The by-pass will restrict the existing irrigation facilities to the land to the east. It will be necessary to provide No 2 300mm diameter sleeves at points to be agreed, under the road to take the irrigation main.

   6.7 Double glaze the windows of the farm residence and cottages and provide any further noise insulation considered necessary.

   6.8 Erect a suitable screen to minimise noise and traffic lights for a distance of 60m run in front of the house.

7. **PLAN**
   The Acquiring Authority to prepare a new farm plan and calculate the area of the fields affected and supply No 5 copies to the claimant.

8. **ADVANCE PAYMENT AND INTEREST**
   An advance payment of 90% of the claim to be made as soon as entry taken.

9. **FEES**
   The Acquiring Authority to pay the claimant's valuers' and solicitors' fees. Advance fees to be paid.

10. **INTEREST**

Interest at the Statutory Rate to be paid on the remaining amount of the claim (after payment of the advance) from the date of entry to date of payment. Such interest to be paid on an annual basis.

E&OE

PRACTICE NOTES

(a) Claim Item No 8 Advance Payment and Interest

The advance payment will be made on the basis of the District Valuer's estimate of the compensation payable and not upon the amount of the claim.

(b) Claim Item No 9 Fees

In view of the huge volume of work involved in claims of this nature, valuers generally require a substantial fee for the work involved. The acquiring authority now pay on the reasonable cost of the work reflecting time, expertise and effort required to do the work. The valuer should keep a note of all time used and other expenses incurred.

STATEMENT OF CLAIM FOR DISTURBANCE

by

Mr JR Yeoman (Freeholder and Occupier)

against

The Department of Transport

as a result of acquisition of part of an entry onto
MOOREND COURT, WESTINGFIELD, ANY COUNTY
to construct the Westingfield By-Pass

March 2008   Prepared by: William Davies & Rees
                                 Agricultural Valuers & Chartered Surveyors
                                 Dolanog
                                 Welshpool, Powys

The Acquiring Authority to pay for the following matters:

1.	The Electricity supply was cut on 14 March 2007, for 18 hours. One milking lost and another delayed and two cows affected by mastitis.	
	Loss of 1,241 litres milk @ 25p	£310.25
	Loss on two cows £300 (one lost a quarter) and vet's bill £70	£370.00
	2 mens' overtime — 6 hours @ £8.50	£51.00

2. Access was denied to the homestead on
   15 separate occasions.
   Inconvenience and loss @ £100 — £1,500.00

3. Cost of additional cleaning of the farmhouse and two
   cottages due to dust and mud over a period of
   50 weeks. 1 hour per day. 350 hours @ £7.00 — £2,450.00

4. Cost of additional cleaning of carpets and curtains
   (receipts attached) — £108.20

5. The farmhouse and two cottages require re-decorating
   externally and internally one year earlier than usual
   due to the works.
   Cost £8,000 over four years, 25% cost — £2,000

6. Access drive was left in such a bad state that farm lorry
   axle broke. Cost of repairs (receipt attached) — £548.75

7. Additional cost of cleaning farm drive arising from the
   work 40 hours @ £8.50 — £340.00

8. Water supply to OS Nos 3, 5 and 7 (east) were cut
   on July 5/6 2007 and 1,620 litres of milk lost @ 25p — £340.00

9. Water supply fracture resulted in 20,000 gallons of
   water being lost. Cost @ 450p per 1,000 gallon — £90.00

10. Dust polluted the grazing in OS Nos 3, 5 and 7 for
    the period of 10 May–28 July (less 2 weeks) 2007.
    Damage to 25 acres. Milk yields suffered and it is
    considered that yields were depressed by
    8,000 litres @ 25p — £2,000.00

11. Dust polluted 2 cuts of silage in 2006 (total
    280 tonnes). It became necessary to feed, on
    Consultants recommendation (copy available)
    additional concentrate and 10 tonnes 16%
    feeding stuffs were purchased from West Central
    Farmers Limited at a cost of £150 per tonne — £1,500.00
    Quantity of silage totally useless.
    Loss 40 tonnes @ £20 — £800.00

12. Precision chop silage harvester damaged by rock blasted.
    Cost of repairs (copy account attached) — £725.40

13. Blasting has caused fracture in south elevation of
    farmhouse. Estimated cost of repairs   £3,000.00

14. Ditches silted up in OS Nos 3/5 and 5/7. Cost of clearing
    out 200 m @ £3   £600.00

15. Part OS Nos 11, 7 (west) and 8 flooded due to
    blockage of stream by contractors. 3.5 acres affected.
    Loss of grazing £50 and silage crop £200   £250.00

16. Damage caused by Contractors entering OS No 7
    (east) in June 2007 to locate and rectify blocked drain   £82.00

17. Damage caused to all enclosures affected when
    contractors entered during wet conditions in November/
    December 2007 to plant quickthorn hedge and remedy
    defective post and wire protective fencing   £170.00

18. Loss of Crops 2007
    Loss of 1 acre Avalon Winter Wheat in OS No 1007
    3 tonnes @ £140 & Straw £20   £440.00
    Additional Cost of Harvesting   £25.00   £465.00

19. Stock got out on six occasions, namely 15, 16 and 18 April;
    20 June, 21 and 28 July 2006
    Cost of rounding up 12 hours @ £8.50   £100.00

20. Damage was caused to the silage crop in OS Nos 8612
    and 7401 (W) on 15, 16 and 18 April 2006
    Loss 3 tonnes @ £20   £60.00

21. To claimant's time wasted arising from the entry. A diary has
    been kept and this shows 41 hours spent by the claimant
    attending meetings with his solicitors, valuers, NFU resident
    engineers, district valuer and considerable time telephoning
    largely to report complaints.
    41 hours @ £20   £820.00
    56 telephone calls @ £1   £56.00
    20 letters written @ £4   £80.00
    161 miles travelling @ 40p   £64.40   £1,020.40

22. Forced Sale of Stock
    The loss of the land and general upheaval caused will
    result in 20 cows having to be sold off, many at the wrong
    time for their optimum sale, ie as stale cows. A capital loss
    arises on this forced sale. 20 cows @ £400   £8,000.00

23. Valuer's Fees & Costs

24. Other Matters Arising

E&OE

Tenant's claim

Assume all facts as given above except that the holding is occupied by Mr John Farmer on an annual Candlemas (2 February) tenancy at a rental of £12,500 pa (revised last on 2 February 2005). The tenant, who is 45 years of age and has two sons of 18 and 16 years of age, holds under a normal agricultural tenancy and repairing liabilities are in accordance with SI 1973 No 1473. He has improved the farm buildings since he took the tenancy in 1994 and has provided most of the modern buildings (these are tenant's improvements executed with unconditional landlord's consent) and all the dairy equipment including the 6/12 herringbone parlour. Because he has carried out these improvements he is enjoying a profit rental of about £35 per ha. The holding has a current investment value of £10,000 per ha and a vacant possession value of £20,000 per ha.

The first action to take in formulating this claim is to agree the rental apportionment between the tenant, landlord and the acquiring authority. This is necessary because the law requires it (section 19 Compulsory Purchase Act 1965) and also to enable the tenant to be paid the Statutory Payment (additional amount) by virtue of the 1968 Act.

Basis of apportionment

This is not a rent reduction under section 33 of the 1986 Act. The apportionment of the rent payable at the date of severance should be on the basis of the value attributable to the severed land and the retained land, respectively, as part of the whole. The rent apportioned to the retained land should have no regard to any depreciation suffered by reason of the severance or the use made of the severed part.

The apportionment of the rent does not operate to create a new tenancy and therefore does not debar the landlord from securing an increase (or the tenant a decrease) on the occasion of the next normal review date in accordance with section 12 of the Agricultural Holdings Act 1986. The effect of any severance will then be reflected in the rental review, and any advantage or disadvantage to landlord and tenant, in the interim, should be reflected in the valuations for the respective compensation claims.

The four heads of claim are set out in 19.3; everything to be claimed must therefore be fitted into one or other of the heads.

STATEMENT OF CLAIM
by
Mr J FARMER (TENANT)

against the Department of Transport
following the acquisition of part of

MOOREND COURT, WESTINGFIELD, ANY COUNTY
for the construction of the Westingfield By-Pass

Prepared by: Williams Davies & Rees
                 Agricultural Valuers & Chartered Surveyors
                 Dolanog
                 Welshpool, Powys

Area Acquired (6.07 ha) forming Part of OS No 1, 2, 3, 5, 7 and 8

The acquiring authority to pay the following claims:

1.   **VALUE OF TENANT'S UNEXPIRED TERM**

1.1 **Value of Tenant's Interest in area taken**
     The tenant is a young man and can be reasonably expected to be farming
     the holding for a further 20 years or so when one of his sons can, provided
     he satisfies legal requirements, be expected to succeed to the tenancy.
     (Agricultural Holdings Act 1986, Part IV, Sections 34–58.)
          Furthermore he has carried out major improvements to the holding.
     Normally the value of the tenant's interest is considered to be 50% of the
     difference between the estimated vacant possession value (£20,000 per ha)
     and the investment value (£10,000 per ha). However, in view of the tenant's
     improvements in this particular case, it is considered that the value here is
     60% of the aforementioned difference

= £10,000 × 60% (6 ha) =	£36.000.00	
Less: 4 × £882 (apportioned rent) =	£3,528.00	£32,472.00
(see also Item 4 of claim)		

1.2 **Licence Area**
     User of the area of 0.6ha as a licence for the duration
     of the contract, ie

2 growing seasons	£1,482.00
Total Value of Tenancy interest	£33,954.00

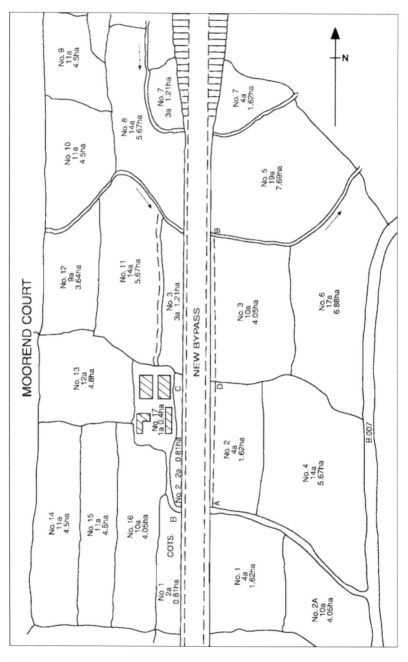

2. OUTGOING TENANT'S PAYMENT

2.1	Unexhausted Manurial Values	£105.00
2.2	Residual value of Feeding Stuffs consumed	£95.00
2.3	Unexhausted value of lime applied (Detailed calculation of 2.1–2.3 attached)	£78.50
2.4	Tenant's pastures in Pt OS Nos 3, 5 and 7 4 ha @ 150	£600.00
2.5	Loss of crops in OS Nos 1, 7 and 8 2 ha @ £400	£800.00
2.6	Labour to Farmyard Manure in OS Nos 3 and 5	£300.00

Total of Tenant Right £1,978.50

3. INJURIOUS AFFECTION TO TENANCY

3.1 Fencing Liability

Injurious affection arises as a result of 1,020m of additional fencing having to be maintained. This will need annual maintenance and when the hedge is of sufficient size (approx. 10 years' growth) it will need laying.
Additional annual expenses including:

(a)	80 hours maintenance @ £8 YP 4%	£640.00 25.00	£16,000.00
(b)	Cost of laying hedge in 10 years' time 1020m @ £12 Defer 10 years @ 8%	£12,240	£5,667.00

3.2 Redundant buildings and equipment

Current value of redundant buildings is
£80,000 — 25% redundancy £20,000.00

Total Injurious Affection £41,667.00

4. AGRICULTURE (MISCELLANEOUS PROVISIONS) ACT 1968 (Section 12)
PLANNING AND COMPULSORY PURCHASE ACT 2004
(Section 33A) (To be calculated)
Occupiers Loss Payment (max £25,000)/Basic Loss Payment (max £75,000)

5. **DISTURBANCE, CROP LOSSES ETC**
   This is to be the subject of an additional claim to be made when the contract is completed [as for the owner/occupier]
   Note: Four times the rent of the affected area will be deducted from the claim (provided the agreed amount of claim is higher than the four times rent; as shown in 1.1, see also Item 4).

6. **ACCOMMODATION WORKS**
   As item 6 of Owner/Occupier Claim.

7. **REDUCTION IN RENTAL**
   A reduced rental of the holding is consequent upon the acquisition to be negotiated and agreed between the Claimant, the Landlord and the Acquiring Authority.

8. **PLAN**
   The Acquiring Authority to prepare a new farm plan and calculate the area of the fields affected and supply No 5 copies to the claimant.

9. **ADVANCE PAYMENT**
   An advance payment of 90% of the claim to be made as soon as entry taken.

10. **FEES**
    The Acquiring Authority to pay the claimant's Valuers' and Solicitors' fees.

11. **INTEREST**
    Interest at the Statutory Rate to be paid on the remaining amount of the claim (after advances) from the date of entry to date of payment. Such interest to be paid on an annual basis.

E&OE

19.7 Claim by landlord ie the reversionary interest

The same general principles of compensation apply to land let as apply to owner-occupied land but it should be noted that under the provisions of section 48 of the 1973 Act, the landlords' interest to be valued is that which actually existed at the date of notice to treat, ignoring the effect on the tenant's security of tenure on the acquiring authorities' proposals or any other compulsory purchase scheme. Any notice to quit served by the landlord, founded upon the authority's scheme, is to be ignored and if the tenant has already left it is to be assumed that he has not done so. The landlord can also make a claim for injurious affection and severance, but it is unlikely in practice that there would properly be any claim for disturbance items, other than surveyor's fees.

Practice Notes (applicable to all claims)

(1)　It is the duty of the claimant's agent to submit a full and detailed claim.

(2)　The district valuer is unlikely to agree to all terms of claim or the amounts claimed. Nevertheless, it is up to the claimant and his agent to substantiate the claims made.

(3)　Frivolous items of claim should not be lodged.

(4)　Each case must be treated individually. Varying matters of claim will arise for different claims and the valuer should not miss items which obviously are valid.

(5)　Where tenancies exist, accommodation works should be agreed between the landlord, tenant and the acquiring authority. Similarly any rental reductions consequent upon the acquisition, but the acquiring authorities are not too concerned on this matter and are reluctant to become involved.

(6)　In appropriate cases the tenant may be able to substantiate a claim for severance although this is usually reflected in the reduction of rent that may be negotiated.

(7)　Note that farm loss payments under the Land Compensation Act 1973 (section 34) are now abolished and replaced by the new Basic Loss and Occupiers Loss Payment introduced by the Planning and Compulsory Purchase Act 2004.

Farm Rents

20.1 Provisions for varying farm rents are contained in section 12 of and schedule 2 to the Agricultural Holdings Act 1986 and are amended by the Regulatory Reform (Agricultural Tenancies) (England & Wales) Order 2006.

20.2 The basic requirement of a notice requiring arbitration to vary a rental is that it has to be served to take effect from the next termination date following the demand for arbitration. On an annual agricultural tenancy this means that a demand for arbitration must be served at least one year prior to the term date. Rents can only be varied every three years. However, it should be noted that some holdings, let on lease, have their own rent revision provisions included and these agreed provisions must be observed especially as some do not necessitate a notice requiring arbitration on the rental to be even served.

20.3 Schedule 2 to the 1986 Act provides that the rental properly payable in respect of a holding shall be the rent at which the holding might reasonably be expected to be let by a prudent and willing landlord to a prudent and willing tenant, taking into account all relevant factors, including (in every case):

(a) the terms of the tenancy (including those related to rent)
(b) the character and situation of the holding including the locality in which it is situated

(c) the productive capacity of the holding and its related earning capacity and

(d) the current level of rents for comparable lettings as determined in accordance with subsection (5).

20.4 Productive capacity means what can be produced from the holding (taking into account fixed equipment and any other available facilities on the holding, eg very good or very poor buildings, irrigation facilities, etc) on the assumption that it is in the occupation of a competent tenant who is practising a system of farming suitable for the holding.

20.5 Related Earning Capacity means the extent to which, in the light of the productive capacity, a competent tenant practising such a farming system could reasonably be expected to profit from farming the holding.

20.6 Para 1(3) of schedule 2 provides that the arbitrator shall take into account evidence of the current level of rents payable or likely to become payable for comparable lettings. This evidence may be cases where rents are fixed by agreement or by arbitration under this Act or are likely to become payable in respect of tenancies of comparable agricultural holdings on terms (other than terms fixing the rent payable) similar to those of the tenancy under consideration. He shall, however, disregard:

(a) the scarcity element, ie the overbid to secure a tenancy

(b) any element of those rents which is due to the fact that the tenant, or a person tendering a rent for a comparable holding, is in occupation of other land in the vicinity so that they may be conveniently occupied together

(c) any effect on those rents due to any allowances or reductions made in consideration of charging premiums.

20.7 Para 2(1) of schedule 2 further instructs the arbitrator to disregard any increase in the rental value of the holding due to:

(a) tenant's improvements or fixed equipment other than those provided by virtue of an obligation imposed on the tenant by the terms of his contract of tenancy

(b) landlord's improvements in so far as he has received government grants in respect of the execution of these improvements.

20.8 Para (3) of schedule 2 further states that the arbitrator:

(a) shall disregard any effect on the rent on the fact that the tenant who is party to the arbitration is in occupation of the holding and

(b) shall not fix a rent at a lower amount by reason of any dilapidation, deterioration or damage to buildings and land caused by the tenant.

20.9 The effect of schedule 2 of the Act is that there are now three very important concepts in farm rental valuations. The arbitrator must:

(a) take into account the productive capacity of the holding and its related earning capacity

(b) take into consideration evidence of the current level of rents for comparable lettings

(c) ignore the scarcity element.

20.10 The productive capacity of the holding

This means basically what a holding is capable of producing in the occupation of a competent tenant practising a system of farming suitable to the holding. Various types of farms exist and some of these types are as follows:

(a) dairy farms

(b) grazing or feeding farms

(c) arable farms — suitable only perhaps for cereal and oil seed rape production

(d) good class arable farms with deep, free-draining, stone-free soil, capable of growing potatoes, sugar beet and similar high value cash crops

(e) rearing farms — generally marginal land holdings
(f) hill farms
(g) specialist farms, eg pig and poultry holdings.

There is, therefore, a wide spectrum of holdings, all which have differing productive capacity. Climate, elevation, aspect, location, etc, all affect the productive capacity as do the quality of soil. The valuer should, however, not place too much reliance on the Agricultural Land Classification Maps. They are of some help but they do not always signify the real quality of the soil concerned, eg most riverside land is shown as Grade III, largely because it is liable to flooding. However, most riverside land is good quality free-draining soil which generally speaking has far greater productive capacity than say some Grade II land and sometimes even Grade I land. Again the productive capacity of fields even on the same farm varies. Some fields have extremely good quality soil which may be capable of producing, in an average season, well over 4 tonnes of cereals per acre (9.8 a ton per ha) while others on the same farm may only produce, on average, say 2 tonnes per acre (5 tonnes per ha). The gross margin difference here on, say, a winter wheat crop may be as much as £140 per acre (£350 per ha). Other areas of a farm, or even a whole farm may be capable of being irrigated, which despite being somewhat costly, can boost yields considerably in drought conditions and thus increase the productive capacity. A contrast is the case of, say, a 300 acre (120ha) Cotswold arable farm with rather shallow stony brash, limestone soil compared with a similar size Herefordshire farm in the Wye Valley with some alluvial silty soil and some free-draining red sandstone soil. Assume both have equal quality buildings including adequate corn storage and good general purpose buildings.

The Herefordshire holding could also, by feeding beet tops and other roots grown, finish off a substantial number of store lambs and thus take advantage of the normal higher prices realisable for them, fat in March when they are sold off.

The differing productive capacity of these two holdings can be illustrated as follows.

120ha Herefordshire farm					
Crop/Ent.	ha	Yield per ha (T)	GM £/ha		Total GM £
Winter wheat	41	8.6	1,050		43,050
Winter barley	25	7.6	860		21,500
Potatoes	8	44.0	800		6,400
200 breeding ewes	38		35	(head)	7,000
60 fattening steers			100	(head)	6,000
Spring barley	8	4.9	350		2,800
	120				86,750
					£722.91 per ha

120ha Cotswold Farm					
Crop/Ent.	ha	Yield per ha (T)	GM £/ha		Total GM £
Winter wheat	25	6.2	1,050		26,250
Winter barley	50	5.4	720		36,000
400 breeding ewes	45		35	(head)	14,000
60 fattening steers			100	(head)	6,000
	120				£82,250
					£685.42 per ha

It will, therefore, be seen that on two similar sized holdings, with substantially the same system of farming, but one with better and more versatile land, capable of producing higher value cash crops, the total GM is better on one by almost £4,500.

20.11 Illustrating further the factors that influence the productive capacity of any holding, reference is made to Chapter 17 with most of the matters referred to therein being relevant. However, to summarise these, the following matters should all be considered by the arbitrator and valuer in determining the productive capacity of any given holding.

(a) Land and Soils

Is it level or is it banky?
Does the land flood?

Are there irrigation facilities?
Is the land versatile, ie suitable for tillage and grazing?
What is the rainfall?
Elevation and aspect?
Is the land well watered, shaded and fenced?
Is the soil free-draining?
Is the soil deep?
Is the soil stony?
Is it early soil or is it cold and late?
Are there wet areas in need of drainage?
Is the soil heavy or is it light?
Can high value crops be grown and harvested?
Is there a high water table?
What is the DEFRA grading of the soil?
Can one winter stock on the land without undue poaching?
Are there any waste areas?

(b) Buildings

What is the standard of buildings and are they suitable and adequate for the type of farming system practised, eg are there good grain storage and drying facilities on a cereal farm and is the dairy set-up suitable and adequate on a dairy farm?

Are the buildings modern and versatile? For example, contrast the existence of a large general purpose building (suitable for almost any use) and the presence of several smaller older type buildings of inadequate size and with too low a headroom.

Are the buildings too specialised in nature and thus restrictive in their possible use?

Are the buildings well designed and of good access?

Are repairing costs likely to be average?

(c) General

Position of the farm — is it convenient or isolated from towns and markets?

What quality of house exists?

Are there sufficient cottages?

Does the farm have a Single Farm Payment (SFP)? It is clear that milk quota has value and can have considerable effect on the earning capacity of any given holding and thus its rental and capital value. Milk quota purchased by the tenant is ignored. If the basic quota has been apportioned between landlord and tenant, the tenant's fraction has no effect in calculating the productive capacity of the holding. Sugar beet quota near a processing factory has value and increases sugar beet earning capacity.

20.12 Related earning capacity

The earning capacity, as the wording of the section states is, of course, related to the productive capacity of the holding, in fact, it directly follows. The definition in the section states that it is the 'extent to which, in the light of that productive capacity, a competent tenant practising a system of farming suitable for the holding, can reasonably be expected to profit from farming the holding'.

This has the partial effect of limiting a rental an arbitrator can award to a reasonable share of the surplus an efficient farmer, carrying on a suitable farming system, can earn. A farmer's accounts may here be of use to a valuer, but the accounts for a single year only can be misleading, for a variety of reasons. If accounts are to be produced, it is considered that to be fair to both landlord and tenant, they should be for three consecutive years, if possible. It should be remembered that sometimes, but not always, accounts are produced for tax purposes and do not always reveal the correct profitability.

Forward budgets can also be of immense value and it might be argued that if produced for one year only, (when prices are known) they may be more valuable than past accounts, which are historic. Both accounts and a forward budget should be produced as they are essential in the rental calculation.

What is a fair share of the net farm income (total gross margin less total fixed costs) that should be paid as rental? Opinions vary — some land agents and valuers consider that it should be 60%. Others, considering that capital a tenant farmer employs (particularly in livestock enterprises) and the somewhat poor returns on the investment, think that 40% is reasonable. This is a matter of contention, but the average percentage that most reasonable minded valuers adopt is 50%, although experience shows that it varies from 40–55%.

Any budget prepared for a rental calculation should be prepared with great care and should provide as much detail as possible as otherwise it could be worthless.

20.13 Current level of rents paid for comparable lettings

Para 1 of schedule 2 requires the arbitrator to take into account any available evidence of rents payable for comparable lettings. This evidence can take the form of any agreed rents or arbitration awards. However, the arbitrator must disregard any element in such rents that is a scarcity element.

The arbitrator must also disregard any element of rent due to the fact that the tenant of any comparable holding is in the occupation of other land in the vicinity, which may be conveniently occupied together with that holding. This simply means that the arbitrator disregards the case of a farmer paying perhaps a very high rent for a nearby area of land (and who can well afford to pay over the odds) since he is effectively spreading his fixed costs over a larger acreage, eg perhaps he can farm such additional holding with the same labour and equipment and thus incur very few additional fixed costs and can afford to pay an excessively high rent.

Evidence of comparable rents, or the general level of rents paid can be very useful to an arbitrator and valuer. However, farms are not like a row of semi-detached houses and direct comparison of farms is well nigh impossible. Quality of land, fixed equipment, approach, situation, elevation, topography, etc dictate that there are rarely direct comparable holdings.

If comparisons are cited, the farm concerned should always be inspected, the tenancy agreement perused, and all necessary information on tenant's improvements, etc obtained. In practice, evidence of comparable lettings must be looked into very thoroughly if it is to be of value.

It will also be worthwhile for an arbitrator and valuer to consider the annual average rental statistics published by DEFRA. However, these must be treated as a guide only but can be very helpful if only to indicate general trends.

Furthermore, any effect on rents due to allowance or reductions made in consideration of the charging of premiums, must be disregarded.

Rentals obtained by letting land by tender under the Agricultural Tenancies Act 1995 may well be cited as evidence. These should be treated with great caution as they are not comparable lettings under the 1986 Act.

20.14 Tenants improvements

Para 2 of schedule 2 states that the arbitrator shall disregard any increase in rental value of the holding due to tenant's improvements or fixed equipment (other than that provided by him as an obligation in his contract of tenancy). It is immaterial whether landlord's consent for such improvements was received or not and the farm must be valued as if the improvements or equipment did not exist. In practice, this is difficult and valuers often value the property as it actually is and then value out such improvements. However, this method should be approached with care as the end result may otherwise well result in over-emphasising the value of the improvements, as part of the whole farm. Alternatively the farm may be valued as unimproved, ie the black patch method where improvements are totally ignored. Three examples of cases of valuing out are set out below.

(i)   10 acres (4.04ha) of very rough grazing reclaimed by the tenant. Previously it had little value (say £12.35ha) but is now good land worth, on the open market, say £99 rent per ha. The calculation is:

Current rental value	£400
Value before improvement	50
Increased net rental value	£350

The net cost of the improvement was £2,500 and provided it is properly farmed, in the future, the improvement will be beneficial to the occupier *ad infinitum*. The increased rental value, charging interest at 10% on £2,500 = £250. This expenditure has therefore released an extra or hidden value (latent value) of £100 pa (£350 − £250 = £100).

    The County Court decision in *Tummon* v *Barclays Bank Trust Co Ltd* [1979] 1 EGLR 14 decided that latent value should belong to the landlord. Logically this is because the landlord owns the raw material for the improvement — here the land (and this could also be argued for the site of a tenant's buildings or other improvement). The landlord should therefore be given credit for the basis on which a tenant derives benefit. In this example the tenant is not entitled to deduct more than £250 pa from the gross rental:

Increased rental value	£350
Less: Latent value	100
Increased net rental value	£250

(ii)   Erection of 3,600 sq ft (335 m^3) covered yard 10 years ago. Total life expectancy is 40 years. Current net costs of the work including site levelling, would be £21,500. Unexpired life is 30 years. Current value is:

£21,500 × 30/40 = £16,125 say £16,000

However, the latent value of the site is say £50 pa (in view of the use now put to the site it is possible that even a higher latent value can be properly justified).

A landlord erecting such a building for his tenant would reasonably expect 10% interest on his investment. A tenant could also expect a similar return. The current net increased rental value resultant on the improvement is therefore:

£16,000 × 10% = £1,600 − £50 (annual latent value)
= £1,550 which sum should therefore be deducted from the gross rental.

(iii) The tenant has divided a 30 acre field into two convenient sized fields by the erection of a 400m length of pig and barbed wire fencing, together with a 3.6m iron gate. The work was carried out three years ago and the current cost of the work would be £900. Total expectation of life is 12 years. The annual value of the improvement is therefore:

£900 × 9/12 years × 10% = £67.50

It is considered here that no latent value has been released and the sum thus deductible from the gross rental value is £67.50 pa.

20.15 Rental valuation

In preparing a rental valuation under section 12 of the 1986 Act, the following course of action is suggested.

(i) A detailed inspection of the holding should be made noting carefully:
  (a) type of fixed equipment, quality of land (which the valuer should carefully assess, grading this himself, into say four categories of differing qualities and carefully assessing its productive capacity and thus its earning capacity having regard to all relevant factors)
  (b) tenant's improvements and fixtures, if any, and calculating their annual value or adopting the black patch approach and completely ignoring them
  (c) how well farmed is the holding and is it being competently farmed?
  (d) is the system of farming practised suitable for the holding?

(ii) Prepare a detailed budget, based on the farming system practised and also taking into account whether the tenant is farming to optimum levels of production on a system suitable for the holding.

(iii) Examine, if possible, farming accounts for the past three years. However, these may not be available or will not be produced.

(iv) Consider all available evidence of rentals paid in the area and possibly, if the farms are comparable, further afield.

(v) Research into the terms of the tenancy. Nowadays many new lettings are on a full repairing and insuring basis which means that any tenant holding on these terms bears a higher burden than a tenant holding on repairing terms equating to those prescribed by SI 1973 No 1473. It is difficult to generalise on the additional burden a tenant takes on a full repairing and insuring basis, but this is likely to add an extra 15% to the norm.

Valuers have differing methods of valuation which often produce not dissimilar results. All the factors previously mentioned above and in Chapter 17 should be taken into account. However, at the end of the day (but taking into account the statutory requirements of schedule 2 to the 1986 Act), what is most important of all is the quality of the soil, its versatility, quality of the fixed equipment and that it is competently farmed to a system suitable for the holding concerned.

Example valuation

Hen Blas, Pengwern, Monmouth, is a 210 acre (85ha) stock rearing/dairy farm in the County of Monmouth with a small modern 3 bed residence, a semi-detached 3 bed workman's cottage and a reasonable set of farm buildings. The land is generally undulating with some severe banks and 95 acres is Grade III and 103 acres is Grade IV land on the Agricultural Land Classification Map. About 12 acres are rough unproductive land and the homestead and access roads. The land lies between 400ft and 700ft above sea level. The valuer has placed the land in his own categories as follows:

A — 94.25 acres (38.14ha) B — 57.75 acres (23.37ha)
C — 44 acres (17.8ha) D — 12 acres (4.85ha)

The farm is not well situated, being 17 miles from the nearest market town and is approached for the last mile by a very narrow council maintained road.

The tenant, aged 44, is a very competent and better than average farmer. He runs the farm with the aid of a 24-year-old workman and some casual labour at busy seasons. 70 milking cows are kept (milk quota is 420,000 litres and it has been informally agreed that 90% belongs to the landlord and 10% to the tenant). Some calves are reared, 15 being dairy replacements and 30 others as beef stores (some Friesian steers, some Friesian heifers and the remainder Hereford cross Friesian cattle) are sold off as 15 month stores. 110 full mouthed Suffolk cross ewes are kept. In view of the type of holding this is, production is considered to be at optimum level and, without intensification and capital investment, cannot be further increased.

The tenant has carried out the following improvements at his own expense:

(a) concreted 210m² around the buildings over the last six years. Net cost £1,560
(b) erected cow kennels for 60 head
(c) erected milking parlour, constructed collecting and dispersing area, together with feeding area (including the provision of mangers) erected dairy.

Note: (b) and (c) was done three years ago at a total net cost of £26,000. Some old buildings belonging to the landlord were demolished on site.

(d) tile drained 21 acres (8.5ha) five years ago at a net cost of £3,850
(e) provided a piped water supply with no 4 concrete water troughs to provide a mains water supply to 56 acres (22.5ha) of the top land. Net cost was £1,210 and the work was done two years ago.

Total net cost of tenants improvements/fixtures £32,620 (£1,957 @ 6%).

The farm is held on an annual Candlemas tenancy with the tenant being responsible for all repairs and payment of insurance premiums. Current rental fixed by arbitration with effect from 2 February 2003 is £7,500 pa.

The best approach to this rental valuation is threefold.

(a) Prepare a detailed budget as shown. It is suggested that in this case 50% of the net farm income (gross margin less fixed costs) should be adopted as a reasonable guide to the rental payable. This budget should reflect the productivity and related earning capacity of the holding.
(b) Prepare a detailed breakdown rental valuation of the holding. This valuation should carefully reflect the quality and value of the house, buildings and land. The valuation should be on a full market rental basis showing the separate gross values of:

  • the residence
  • cottages (if any)
  • buildings
  • the four categories of land (or less if there are lesser categories).

This will produce the gross rental value from which must be deducted the calculated annual rental value of the improvements, or to totally ignore these improvements as if there was a black patch over them. It is also suggested that if the holding suffers from some severe limiting factors (which have not been taken into account in the gross rental value computation) such as poor approaches, isolation, an awkwardly shaped farm and other constraints, percentage deductions of between 1 and 10% should also be deducted from the gross value, as should any undue repairing and insuring burdens. If, on the other hand, the farm has some outstanding features, not already accounted for, such as good position, irrigation facilities, etc, there should be added a suitable percentage to the gross value to reflect these. Also any other special advantages such as the inclusion of sporting rights in the tenancy should be valued.

The result should produce a fair and balanced rental valuation.

(c) Consider any evidence of comparable lettings. The exact terms of any comparison should be ascertained and suitable adjustments made if anomalies exist between these and the subject land. Also adjustments made to reflect quality of land, scarcity element, etc.

FARM BUDGET APPROACH

All prices based on current values

1. **Land**

Farmable land	198 acres	(80.16ha)
Rough, Buildings etc	12 acres	(4.84ha)
Total	210 acres	(85.00ha)

2. **Enterprises**

Dairying

70 dairy cows @ 1.4/acre/head (0.56/ha) head	98 acres	(39.6ha)
15 dairy replacements @ 2.5/acre/head (1/ha) head	37 acres	(15.0ha)
(1 unit = 1 calf, 1 yearling and 1 heifer)		
30 beef cattle @ 1.2/acre/head (0.48ha/head)	36 acres	(14.5ha)
110 ewes @ 4/head/acre (0.1 ha/head)	27 acres	(10.9 ha)
Total	198 acres	

3. **Gross Margin Summary**

70 dairy cows @ £847 head	£59,290
15 dairy replacement units @ £460/unit	6,900
24 head store cattle @ £80/head	1,920
110 ewes @ £24.77/head	2,774
Total gross margin	£70,884

4. **Fixed Costs**

Labour		
1 full-time man (inc. overtime)	16,000	
Machinery Costs inc:		
Depreciation, repairs fuel and oil, contract,		
sundries £95/acre (£234/ha)	19,950	
Water and Electricity	3,500	
Miscellaneous inc:		
Telephone, rates, insurance, general		
maintenance, sundries	9,500	
Total Fixed Costs	£48,950	

5. **Net Farm Income**

(Gross margin less Fixed Costs)		
Gross Margin	70,884	
Less Fixed Costs	48,950	
		21,934
Add Single Farm Payment	6,200	28,134
Deduct for Tenant's Improvements		1,957
Net Farm Income		26,177
50% of NFI as rent		13,088
Net Rent Payable		£13,088

GROSS MARGIN ANALYSIS

1. **Dairy Cows**
   (70 head — Quota 420,000 litres = 6,000/cow)

Output		
6,000 litres @ 26p	1,560	
Calf (allowing for losses)	86	
	1,474	
Less: Cow depreciation	77	1,397
Variable Costs		
Concentrates 1.67/tonne @ 168/tonne	280	
Vet and Medicines	51	
AI and Recording	29	
Sundries and Bedding	70	
Forage costs	120	550
Gross Margin per cow		£847

2. **Dairy Replacements**
Each dairy replacement unit represents 1 Calf, 1 Yearling and 1 Heifer

Value of down calving heifer	900	
Less Calf value	85	
Net output		815
Variable Costs		
Concentrates inc. calf food	170	
Vet and Medicines	35	
Bedding	40	
Forage Costs	110	355
Gross Margin per head		£460

3. **Store Cattle**
30 head are kept and selling at 12 months of age

Output		
Sale of 385kg cattle @ 115p/kg	442	
Less: Cost of Calf	130	
Net output		£312
Variable Costs		
Concentrates and bulk feed	140	
Vet and Medicines	22	
Miscellaneous Bedding, Forage, etc	70	232
Gross Margin/Head		£80

4. **Sheep**
Lambing 1.5%

Output		
1.5 lambs × 40kg @ 106p/kg	63.60	
Wool	1.25	64.85
Less: Ewe & Ram Depreciation		£13.10
		£51.75
Variable Costs		
Concentrates	8.08	
Vet and Medicines	4.90	
Sundries and Forage	14.00	26.98
Gross Margin/Ewe		£24.77

RENTAL VALUATION APPROACH

94.5 acres Class 'A' land @ £50	4,725	
57.75 acres Class 'B' land @ £35	2,021	
44 acres Class 'C' land @ £30	1,320	
14 acres Class 'D' land namely, 12 acres rough @ £5	60	
House	2,600	
Cottage	1,560	
Buildings	5,200	
Gross Rental		£17,486

210 acres Total

Less Deductions		
Tenant's Improvements/Fixtures, 6% of £32,620		1,957
Repairs	3,000	
Insurance	2,000	
		5,000
Disability of holding		
Isolated and a poor shape with considerable uphill pull to the homestead, which is badly placed in relation to the land and with banky land		
Deduct 3% of gross rent of £17,486		525
Total deduction		7,482
Rental value		£10,004

Summary, the rental should be between £13,088 and £10,004 = £11,546, say £11,500 (£54.76 jper acre, £135.29 per ha).

Valuation Clauses on Sales of Farms

21.1 Differing valuation clauses are included as part of the conditions of sale of farms and estates, and their extent is largely dependent on where the property sold is situated. Agents operating in certain areas — particularly the south of England — have wide, all-embracing clauses, often covering the full extent of a valuation when a tenant quits. It is thought, by some, that it is unreasonable for a vendor to expect a purchaser to pay a full tenant right type of valuation in addition to the price for the freehold.

21.2 Care must be taken in drafting valuation clauses in order to avoid subsequent disputes. The basis of the valuation should be clearly given. It is usually prudent to stipulate that hay and straw should be valued on a market value basis. Consuming value usually has no relevance in a sale of a farm since more often than not it is the vendor vacating, and not a tenant. However, should it be necessary to value hay and straw at consuming value, reference should always be made to the consuming value prices fixed by the local branch of the CAAV, as this will be a firm basis for the valuation. It is best to stipulate, where the silage clamp is unopened at the date of completion, it should be taken over at a fixed sum. If the clamp is opened and in use, it is advisable to state that the valuation should be made on the basis prescribed by numbered publication 183 of the Central Association of Agricultural Valuers. Even then, a disagreement on value is a distinct possibility.

21.3 If crops are mature at the date of completion, it is advisable that these should be valued on a crop value basis, ie the value of the mature crop — but to avoid dispute, whether mature or not, it is best, wherever possible, to stipulate a sum to be paid for such crop.

This, of course, can include enhancement value, if it is thought to be applicable.

Where crops are recently sown and where cultivations carried out, it should be clearly stipulated that these are to be taken over at cost of seeds, fertilisers, sprays, and the cultivations are to be paid for at CAAV rates or contractor's costs (where they have been carried out by a contractor) and also enhancement value.

21.4 Where young seeds are to be paid for, the sum to be paid for them should be given.

21.5 Where equipment or fixtures are to be paid for, again to avoid dispute, the sum to be paid for them should be given.

21.6 In recent years it has become noticeable that the settlement of valuations is becoming an ever-increasingly long drawn out process — possibly, in some cases, to delay payment. This is unfair to the vendor, and it is thought advisable that, unless a sum is paid on account, certified by the outgoing vendor's valuer, on completion date, the balance of the finalised amount of the valuation or any arbitration award thereof should carry interest from the date of completion to the date of payment at, say, 4% above the current bank base rate.

21.7 A clause stating that no claim for dilapidations, deterioration of any other offset should be included. This is most important, and particularly so where a farm or land is sold which was previously occupied by a tenant.

21.8 It is important that a valuation clause should be clear, specific and not ambiguous in any way, as disputes will otherwise arise. It is not sufficient to say that the valuation should be carried out in the usual way. The valuation clause should state that, where applicable, valuers are to be appointed by each party and in the event of disagreement, the matter settled by an arbitrator appointed by them or, failing agreement, by the President of the CAAV.

21.9 A common pitfall is for the term tenant right to be used in referring to items to be taken to. Tenant right has no place in vendor and purchaser relationships and this reference should be avoided. If, for instance, unexhausted manurial values, unexhausted lime and residual values of feeding stuffs consumed, tenant's pastures, etc, are to be paid for, the valuation clause should clearly and specifically state what is to be taken too and also on what basis the valuation is to be calculated. Alternatively, a price per acre should be given.

21.10 A suggested specimen valuation clause is set out below.

The purchaser shall in addition to the purchase price pay the sum of £. . . . . . for the fitted carpets in the lounge, hall and staircase of the farmhouse, the young seeds sown in OS No 123 and the Desco 1,200 litre refrigerated milk tank and Alfa-Laval milking machine with Brooks 3 hp electric motor, and shall also take over and pay for at valuation on completion, the following:

(a) all hay and straw remaining on the holding, at completion, at market value (alternatively)

(a) all hay and straw remaining on the holding, at completion, at consuming value as fixed by the . . . . and District Agricultural Valuer's Association)

(b) silage remaining on the holding, at completion, valued in accordance with the basis prescribed in numbered publication 183 of the CAAV

(c) the growing crop of winter barley in OS No 456(ha) at the cost of cultivations, seeds, fertiliser and sprays, together with enhancement value of £ (per acre or £ per ha)

(d) all cultivations and acts of husbandry carried out, fertilisers and sprays applied to the unplanted arable land since the last crop was harvested

(e) labour to farmyard manure to heap in OS No 2789.

All acts of husbandry to be valued at CAAV costings or contractor's costs, whichever applicable.

No claim will be made for the valuable unexhausted value of fertilisers, or lime applied or the residual manurial value of feeding stuffs consumed, and there shall be no counterclaim for dilapidations (if any), deterioration or other offset made whatever.

The valuation is to be made by two valuers, one appointed by each party, and, failing agreement, shall be referred to a single arbitrator, appointed by the valuers or, in the event of disagreement, by the President of the Central

Association of Agricultural Valuers. Such reference will be under the provisions of the Arbitration Act 1996.

If the amount of the valuation has not been agreed by completion date, then the purchaser shall pay to the vendor's solicitors, as agents for the vendor on completion date, such sum as shall be certified by the vendor's valuer, as payment on account of the valuation, the remaining balance to be paid within seven days of such valuation being determined or agreed. If full payment is not made upon completion the balance will carry interest at 4% above . . . . . . Bank plc's base rate from date of completion to date of payment.

As an alternative, it may be preferable, dependent on the size of the valuation, to substitute the following clause:

If the valuation is not settled and paid on completion date, the purchaser shall pay interest at the rate of 4% above . . . . . . Bank plc's base rate on the sum finally determined due from date of completion to date of payment.

Practice points

(i)   It is advisable to keep valuations as simple as possible. It has been found that excessive over-burdening valuations frequently have a depressing effect on prices obtained for the freehold.

(ii)   Do not include petty items in a valuation.

(iii)   Word the valuation clause very clearly, always giving the basis of valuations.

(iv)   Do not refer to tenant right where obviously this is out of context and has no legal standing.

(v)   If unexhausted manurial values, unexpired value of lime and residual value of feeding stuffs are to be paid for, do not refer to them as tenant right — they are actually improvements.

(vi)   If labour to farmyard manure is to be charged for, only charge for this where this is a haul of some distance from the buildings. It is unfair to charge where the heap is almost alongside or very near to the building to be cleaned out — surely, is it not the duty of a vendor to clean out his buildings?

Records of Condition

22.1 Many land agents and valuers have in the past dismissed records of condition as of little value. Nevertheless, section 22 of the Agricultural Holdings Act 1986 provides for such records to be made at the request of either party at any time, and also the tenant can request that a record of improvements and tenant's fixtures should be prepared at any time during the tenancy. If either party does not agree that a record should be made, the President of the Royal Institution of Chartered Surveyors, on request, can appoint a person to make such record. Unless there is agreement on costs (and it is usual for the person requesting the record to be responsible for costs) section 22(3) states that the costs of the record in default of agreement, shall be borne in equal shares by the parties.

22.2 A record of condition is often a very valuable document more especially to a tenant when he is vacating and there is a possibility of a claim for dilapidations arising. It is also very useful in partnership farming, where the partners agree to keep a holding farmed to the standard it was when the partnership commenced.

22.3 Nevertheless, a record is only of value if it is a good detailed document which deals specifically with all aspects of the subject property. Many records prepared in the past have been altogether too skimpy and too general. A tenant has a duty to farm well, and should he take over a run-down holding he would, in normal circumstances, be expected to put the place in order in, say, five years. Even so, a record would be of value in this case, as it would identify all fixed equipment and its condition together with all aspects of the condition of the land including state of ditches, culverts, outfall drains, fences, hedges and also the condition of the land.

22.4 Another example of the value of a record is where a farm has old, wide, very gappy, high hedgerows, which were not stockproof at the commencement of the tenancy, or even hedgerows which though shown on the OS map as existing, do not in fact exist. Cases have arisen where claims have been submitted for replacing gates and the missing hedgerows, or making stockproof hedgerows which were no more than a few clumps of isolated thorn and trees. Again, any nails or staples in trees can be recorded and, thus, claims avoided. The presence of certain weeds that cannot be satisfactorily and easily eradicated, eg blackgrass, can also be recorded.

22.5 A record should always be clear, precise, concise and easily followed. A scale plan should be included with OS numbers and areas of each field shown. In most cases, photographs should be used and properly identified in the record.

22.6 It is customary for a record prepared by a surveyor for a landlord or tenant to be checked and possibly amended by his opposite number, and for both parties to sign the engrossed record as an agreed document. This certainly has the effect of removing any doubt as to the accuracy of the record. Two copies are always provided, and these should be attached to the tenancy agreement.

Example

RECORD OF THE STATE AND CONDITION
of the holding known as
FLATMEAD PARK LAND
FLATMEAD
in the county of Monmouth.

Landlord: HJ Scrivens Esq
Tenant: Flatmead Farming Partnership
Record taken and made: 26 June 2007

NOTES:
(a) Unless otherwise stated in this record, all items are deemed to be in a reasonable and tenantable order.
(b) The plan attached should be referred to for the location of some of the references made below.
(c) Photographs in Annex 1 refer to the marginal numbers shown in the Record.

LAND
1. **OS No Pt 1297, Pt 1304, 1305 and 1306**
   Flatmead Park. 104.498 acres.

1.1 Description
This field forms Flatmead Park and is open level parkland (now cultivated) with No 21 oak trees and 9 horse-chestnut trees growing. Note that the original boundaries between OS Pt 1297 and OS Pt 1304, 1305 and 1306 have all been removed and boundaries levelled.

1.2 Cultivation
Winter barley.
Electricity transmission line traverses in a NW–SE direction.

1.3 Weed infestation
There is a severe blackgrass infestation throughout this field and wild oats are also widespread.
There is sporadic infestation of couch, charlock, groundsel, redshank and other annual weeds throughout the field.
At the headlands there are patches of infestation of nettles, docks, fat hen, thistles, broad-leaved plantain and hogweed.
There is also a widespread infestation of all these weeds plus sterile broome under all the parkland trees.

1.4 Drainage
There are several wet areas in the field, indicating either absence of drainage or a breakdown in any drainage system existing. No drain outlets were found in the ditches.

1.5 Boundaries
1.5.1 OS Pt 297/Road — east
Low hedge trimmed last season, but now overgrown and in need of trimming. This hedge is gappy, weak and is not stockproof.
There is an infestation of cleavers, sterile broome and hogweed in this hedge.
General encroachment of thistle, docks, fat hen and nettles at headlands up to 3m into field.
Ditch (on roadside side) overgrown and in need of cleaning out.

Single-strand barbed wire protective fence in reasonable order but No 10 loose posts.

1.5.2 OS Pt 1297/Road — north east

Low hedge overgrown in places and in need of trimming. Hedge infested with weeds, particularly cleavers and nettles. There are several gaps and also six sections which, in the past, have been fenced with post and 4-rail (offcuts) fencing and which is now weak and defective.

General encroachment of fat hen, nettles, thistles and brambles up to 2.5m in places at the headlands.

Ditch (on roadside) very overgrown and filled in in places and needs cleaning out. No protective fencing.

Gateway with painted metal gate and iron posts which are rusting and the top rail is buckled. Gate does not hang properly and has no latch. Hanging post loose. Gateway overgrown. Four timber side rails — all defective.

1.5.3 OS Pt 1207/1308

Boundary belongs away.

Severe infestation of nettles, docks, thistles, blackgrass and sterile broome up to 10m at headlands into field.

Ditch at side is very overgrown, not draining, and in need of cleaning out.

Two-strand barbed wire protective fencing in very poor order.

1.5.4 OS Pt 279/Road — north west

High overgrown hedge which is encroaching approx 2m into the fields. Hedge is weak at base and gappy. Two large gaps adjacent to two hedgerow oak trees. Hedge in need of cutting and laying. Not stockproof.

Ditch (on roadside) overgrown and fallen in in places, and in need of cleaning out. No protective fencing.

No 2 oak trees have barbed wire stapled into them.

Painted metal gate bent and rusty, and does not hang properly.

No 5 timber side rails badly split.

An area of scrub woodland in NW corner overgrown.

BUILDINGS

2. **OS No Pt 1501 — 2.04 acres**

   Description

   Pt OS No 1507 contains the buildings used by the farming partnership.

   These buildings have been in existence for a number of years and are consequently ageing and wearing. They comprise:

2.1 **Grain Store 90ft × 60ft**

   Constructed of steel, portal framed with Big Six asbestos roof, corrugated iron side cladding, at N end, E side and part S side. Concrete floor. Timber grain walls are in very poor and damaged condition, with timber boarding generally damaged and defective, and 48ft run is missing. Corrugated iron gable end at S side damaged and rusting. Steelwork all rusting and paint peeling off. Stanchion at SW end damaged by impact and buckled. The asbestos roof is weathered and lichen is growing on extensive areas of the east side. Half-round guttering uneven and has no fall, and all joints leaking. Stop end missing at NE end. All gutters need cleaning out. Asbestos rain-water downpipe at SW has bottom section cracked and shoe missing. Concrete floor uneven and wearing, with some cracks evident.

2.2 Lean-to Implement Shed (attached to Grain Store)

Five-bay open-sided steel and Big Six asbestos roofed Implement Shed with part hardcore and part earth floor. Corrugated iron gables. Steelwork rusting, as is the corrugated iron. Asbestos guttering in similar condition to Grain Store but both rainwater downpipes missing.

2.3 Fertiliser Store

Constructed of concrete block walling with asbestos roof. Single steel framed 6-pane window and ledged, braced and battened door.
Concrete floor.
Serious vertical settlement crack the length of the E wall. No 2 window panes broken and No 1 cracked. Window rusty. Door handle missing, and door secured with hasp, staple and padlock.
Door damaged at base.
All paint work in poor order.
Crack in concrete floor.

2.4 Yard and Access Road

Hardcored road and yard. Very uneven surface with many depressions and several potholes, with a cluster of these at the entrance. Ditch alongside road overgrown and blocked. No protective fencing. Pair of 10ft metal entrance gates, which do not shut properly and need re-hanging. Side rails on E side missing. Fence against OS No 1561 belongs away.

We, the undersigned, having inspected the holding DECLARE the above to be a true and accurate record of the condition of the same as at the date taken.

. . . .   BSc, FRICS, FAAV
Chartered Surveyor
Acting on behalf of HJ Scrivens Esq, Landlord

. . . .   FRICS, FAAV
Chartered Surveyor
Acting on behalf of Flatmead Farming Partnership, Tenants

Note: Photographic records should also be identified, initialled and signed, as being accurate.

Livestock

23.1 Valuation of livestock is a complex subject, and it is accepted by those in the livestock industry that some people have natural ability, whereas others have not. Nevertheless, constant handling of stock in markets and noting types, weight and prices will teach most people the rudiments of livestock valuation, even if they do not possess natural flair. It is important to have a thorough knowledge of the class of stock being valued, as there is a world of difference between valuing, say, Friesian dairy cattle and beef stores. Many of the larger firms of livestock auctioneers have specialists in their own fields to deal with different classes of stock. This chapter can only touch upon the rudiments of the subject.

23.2 Beef cattle

Currently, the most popular beef cattle in demand are the continental crosses, ie progeny of native breed cows bulled with continental breeds, perhaps in this order of merit: Limousin, Charolais, Simmental, Blonde D'Aquitaine, Belgian Blue and Marchigiana. The reason is because these crosses produce well fleshed carcases, often of good conformation, and produce a good percentage of saleable lean meat. However, it should be noted that these continental type cattle spend a long period of their lives growing, and as a rule do not fatten off grass unless they are fed concentrate or are out of pure beef cows, such as Herefords, Angus or Welsh Blacks. Pure Friesian steers are still very popular, provided they are out of good Friesian stock (now seldom seen) and do not have much Holstein blood. Holstein dairy cattle have become very popular in recent years, but they are a large, rangy, lean type of cattle, which have generally poor conformation and for beef production there is much difficulty experienced by beef producers in getting these cattle finished. Native breeds such as Hereford, Angus, Longhorn and South Devon are now commanding a premium as the meat is generally tastier than that produced by continentals.

Hereford steers meet a good demand and sell very well, but Heifers are not popular because they are too small and carcasses tend to be very fat. Aberdeen Angus cattle have now returned in popularity and command a premium particularly if pure bred. The beef produced from these pure beef breeds — particularly steer beef — is in good demand but in many cases tends to be rather fat, especially heifer beef.

The ideal beef animal has a body shaped like a brick — well-fleshed, firm, has good length with well developed hind quarters and loins, where the high priced cuts are found. It should not have a large belly.

It is a fairly easy task to a livestock auctioneer, farmer, butcher or dealer who is well versed in his job to value beef cattle. A quick assessment is made of the breed of animal, its conformation, its condition and its age. The valuer will assess the weight in kilograms (unless the animal has been weighed as in a market) and that is multiplied by the current price per kilo obtainable for that class of animal.

23.3 Dairy cattle

It is accepted in the profession that valuing dairy cattle is a far more difficult task than valuing beef cattle. The modern Holstein Friesian has become, universally, the most popular dairy animal in the UK. The Ayrshire and Dairy Shorthorns, once popular, have largely disappeared from whole counties. The Channel Island breeds — Jersey and Guernsey — are not popular with many because of the poor bull calves they produce, and also their limited cull value.

In valuing dairy cattle, the following matters need consideration.

(a) Breed.
(b) Whether pedigree or not. If so, is it registered with the Breed Society?
(c) If pedigree — what is the breeding? Some blood lines are extremely popular, particularly cows sired by well-known proven

bulls. Most pedigree breeders are fully aware of the most popular blood lines.

(d) Age? Old cows with a limited productive life are not wanted unless they have a good pedigree or are, perhaps, in calf to a well-known bull and there is a possibility of having a heifer calf born.

(e) If recorded or not? Good records of milk produced enhance the value of a cow. Yields of over 6,000 litres in a normal lactation (say 305 days) are considered good. Low cell count is also very important.

(f) The milk is of good quality. Milk is now paid on compositional quality, based on fat and protein. (Average quality milk has 4.1% fat, 3.3% protein.) Also, milk produced in late summer/early autumn is more valuable than that produced in the peak grass period of May/June. This means that a producer selling high quality milk at the right period will get more than the norm for his milk and this obviously will be reflected in the value of dairy cows that meet this criteria.

(g) The cow is good on inspection. A good cow should have dairy-like qualities (contrast a very blocky beef type cow which rarely gives much milk) and should be in the shape of three triangles or wedges — looking from the rear end, sideways and from above. She should have a well formed spacious udder — stretching forward underneath and up towards the tail at the end. Teats should be well spaced, and there should be no spare teats. The teats should be free from warts. It should have good milk veins curling forward from the udder. The cow should be well marked and true to breed type.

(h) The animal must be healthy. Many cows have lost quarters (due possibly to mastitis), have cut teats or perhaps, if old, large udders almost touching the ground. Any defect of this kind reduces the value appreciably.

(i) It must have good conformation so that it will realise a good price when sold as a cull cow. Ayrshire and Channel Island breeds have lost popularity because they do not realise high prices as cull cows and do not produce good beef cross calves by comparison with Friesians.

(j) What stage the cow is in its lactation and, if served, when it is due to calve. Cows laying off calving a long time (stale) are obviously not as valuable as a cow that is due to calve or is freshly calved.

(k) Cows over three lactations are not sought because of high cell count.

Having taken the above factors into count, valuations can only be made by direct comparison with other cows or herds sold. Even when whole herds are sold, there is great variation in values — one might find a good young commercial cow, just calved, realising say 1,500 guineas, while a few minutes later an older stale cow, not of particular merit, may only realise say 350 guineas.

23.4 Definitions, etc

A slip or cow heifer is a heifer that has aborted a calf. It will have developed an udder.

A bullock or steer is male bovine, castrated. This is usually done with a bloodless castrator at the age of two to three months.

A three quarter cow or heifer produces milk from three quarters only — the other being blind.

Age of cattle — cattle have baby teeth until they are 21 months old. Two larger teeth then appear, followed by, for each year afterwards, two large teeth (permanent incisors) until they have eight large teeth at four years old.

Store cattle are generally sold by their producers to others for growing further, and often finishing as beef animals.

Dual purpose cattle are cattle kept both for milk and beef, eg South Devon, Dairy Shorthorn and Red Poll.

23.5 Sheep

This chapter does not deal with pedigree sheep, most flocks of which are kept for the purpose of producing ram lambs, or rams for sale, and also some ewes, or ewe lambs for sale to other breeders, who will buy them for breeding purposes.

The great majority of flocks kept are commercial flocks, kept for fat lamb production (in some cases production of store lambs) with production of wool a secondary matter (the price of wool has been very low for years).

The sheep industry has been revolutionised in the UK in the last decade or so. Many formerly popular breeds have lost their popularity because they no longer produce the type of fat lambs required by the meat trade and, in particular, meat exporters. The demand today is for a well-fleshed lean carcase, the most popular weight being from 37 to 42kg live weight. Very heavy fatty carcases are no longer wanted because the modern housewife does not want fat, and this has to be trimmed off at a considerable loss to the retailer. Many of the breeds producing heavy fat lambs have thus declined, in particular the pure bred Down, Clun Forest, Kerry and other similar breeds. Nevertheless, they are still used by some flock masters as crossing breeds and, as such, produce a ewe that is in demand, eg the English halfbred, which is the progeny of the cross of a Blue-faced Leicester ram on a Clun Forest ewe.

The Suffolk cross is still arguably, to some butchers, one of the most popular lambs, since it is usually fat at the weight the trade demands and also has good conformation, a small head and fine legs. It does not get too fat and has little waste on the carcase. The most popular breeds are the Texel and Charolais crosses.

Other very popular breeds are crosses, such as Mule (Blue-faced Leicester ram × Swaledale ewe), Lleyn, Welsh halfbred (Border Leicester ram × Welsh ewe), Welsh Mule (Blue-faced Leicester ram × Welsh ewe or Brecon Cheviot ewe), Scotch halfbred (Border Leicester ram × Cheviot ewe). However, Mules are not always popular as some have poor, narrow hindquarters.

There are also variations which are popular in certain areas, such as the Scotch halfbred (Border Leicester × Cheviot) which is in much demand in the Midland Counties — but the lambs tend to be of heavier weights at maturity.

Valuation of store lambs is a matter of experience, taking into account breed, condition, weight and the trend of the market at the time of valuation. Finished lambs can be fairly easily valued if the correct weight can be estimated and by reference to market prices.

Valuing breeding ewes is more difficult. The following factors have to be considered.

(a) Breed.
(b) Type.
(c) Age. Ewes have only bottom teeth. Up to 12 months a ewe has its suckling teeth. At one year it has two large teeth, at two years — four, at three years — six, and at four years — eight. It is then called full mouth and can exist as such for some time. Eventually, it loses teeth and then becomes broken mouth meaning that it is aged and of much less value, possibly of only cull value. However, an experienced person can tell if a ewe is old, despite having a so-called good mouth.
(d) Udder. It is most important that a ewe should have a good udder,

as otherwise it cannot rear its lambs well. The udder should be sound and when a ewe lambs, should be full of milk.

(e) Health. The condition of a ewe is important. To produce and rear a lamb well, it must be in very good health. Some flockmasters lose lambs (and ewes) at, or near, lambing time due to pregnancy toxaemia (so-called twin lamb disease). This arises as a result of the pregnant ewe not being fed well enough to sustain herself and the lamb growing inside her. It is, therefore, important that pregnant ewes should have supplementary feed for at least a month prior to lambing (say 0.25–1kg ewe cobs a day). After lambing, many flockmasters creep feed their lambs for several weeks if early lambs are to be produced. In any event, ewes should be fed hay and concentrate, both before and after lambing, as grass is usually short at this time of the year.

Valuations of ewes are made by direct comparison with the realisation prices of comparable ewes, taking the above into account.

Notes

(i) Ewes are usually injected about two to three weeks prior to lambing, against Dysentery, Pulpy Kidney, Blackleg, etc. This results in both ewe and lamb being protected.

(ii) Young male lambs are castrated (usually by the rubber ring method) at birth. Some flockmasters also cut the end of their tails at the same time.

(iii) Unless sold fat off their mothers, lambs are generally weaned at about 12–14 weeks of age.

(iv) Ewes and lambs sold together are called couples.

(v) Fat sheep are sold in £ and p per head. Prior to sale time, they are inspected by a grader, in the market, and provided they satisfy his requirements, and are properly finished, ie not too thin or too fat, the live weight in kilograms is announced at the time of sale.

23.6 Pigs

This section is written by Alan N Lane FRICS, FAAV. It deals with the valuation of commercial pigs as opposed to pedigree pigs. Just as pig keeping has become more of a specialised, intensive system of farming over the years, so has the valuation of this class of stock. The number and proportion of store and slaughter pigs sold through livestock markets is sadly declining and therefore the opportunity for young valuers to see pigs sold by auction is declining.

As with beef and lamb, the housewife's demand for lean meat and the farmer's demand for a quick growth rate has resulted in the rise in popularity of certain pure breeds and hybrids. Imported double muscle breeds such as Pietrain pigs are being introduced for cross breeding. The decline in popularity of some of the older breeds used in years gone by to produce big, and often fat, carcases, eg Gloucester Old Spot, Tamworth and Berkshire has been partially stemmed by demand from rare breed survival members and those selling comparatively small quantities of meat at farmers' markets. An example of traditional pork breeds remaining popular are Large White, the Welsh and the Landrace.

The butcher is looking for a long carcase to produce pork chops and a rounded hind quarters to produce legs of pork and hams, and this is achieved often by the crossing of popular breeds. Feeding is also an important factor in achieving the desired end product.

The valuer will often encounter three classes of pigs.

(a) Breeding stock. This includes breeding gilts, sows and boars. With a breeding sow or gilt, conformation is important but also the general health and well-being. She should be well fleshed but not excessively fat and have good feet and legs. She should have an adequate number of well placed teats — 14 is ideal. There should be no sign of mastitis or other infection.

A stock boar will usually be of pure breed (often a Large White) and again must have good general health, sound feet and legs and not be fat and lazy.

Important factors affecting valuation of sows or gilts in farrow are the farrowing date, the number of litters she has produced and the boar by which she has been served. The average sow produces six litters in a lifetime, so when valuing a sow who has had, say, four or five litters her likely cull value is an important factor in establishing a value. With a young gilt or sow, her potential to produce another four or five litters must be regarded.

Often a breeding sow will be valued with her litter of pigs which may be up to six or eight weeks old. They are valued as a family, the overall value depending on the condition and size of the sow, the number, evenness of size and general health of the piglets.

(b) Store pigs. This description covers the range from weaned piglets up to approximately 60kg liveweight. Pigs are weaned generally at six to eight weeks old (although some intensive systems wean at 21 days old) and are commonly referred to as weaners up to the age of 10 to 12 weeks old when they may weigh up to 30kg liveweight. Factors affecting valuation include breed, size, conformation and general health and vigour. Small pigs intensively reared and spending most, if not all, of their life inside, are particularly susceptible to disease although modern veterinary techniques and environmentally controlled buildings reduce this to a minimum. Often piglets' tails are cut at birth. This is a plus factor when valuing store pigs as risk of tailbiting causing consequent stress and injury is reduced. Male piglets are now rarely castrated. Valuation is usually by direct comparison.

Some feed compound firms operate weaner groups organising the direct movement of weaners from the breeder to the fattener. Often these pigs are bought unseen by the buyer who probably takes regular supplies from one or two farms and who relies on them being of a consistent standard. These pigs often change hands at a payment based on the gross weight of the load of pigs or a fixed rate per head plus an additional charge based on weight.

(c) Fattening pigs. Pigs upwards of 60kg liveweight being fed for slaughter are usually referred to collectively as fattening or prime pigs. Weight range descriptions to record prices achieved for fatstock market report purposes are: pork pigs, 60–75kg; cutter pigs, 76–85kg; bacon pigs, 85–104kg; overweights, 105 kg+.

Fattened pigs sold liveweight are weighed and sold in pence per kilogram liveweight, consequently the valuation of this class of stock is often by visual assessment of weight multiplied by a value in pence per kilogram. Factors affecting the value per kilogram include the weight (porkers and cutters are usually worth more per kilo than baconweight pigs), sex (gilts tend to command a higher price than boars), leanness and conformation (which will dictate the all important killing-out percentage).

Generally the selling weight of prime pigs has increased in recent years and there have been significant genetic and nutritional improvements in order to produce a uniform, lean finished product.

Arbitration and Independent Expert Determination

24.1 Many agricultural valuers will sooner or later experience cases that cannot be settled and in order to resolve the matter arbitration will be necessary. The law on arbitration is fairly extensive and there are many legal cases, decided, which are relevant. However, this chapter is intended to be a brief guide of the practical application of the law.

24.2 Arbitration was intended to be a relatively cheap method of solving such disputes but in recent years has proved to be in some cases a somewhat more expensive method of deciding issues than originally intended. It is therefore very important, in the clients' interest, that every effort should be made to settle valuations and disputes without resorting to arbitration.

24.3 Under the provisions of the Regulatory Reform (Agricultural Tenancies) (England & Wales) Order 2006 all disputes relating to 1986 Act tenancies are now dealt with under the provisions of the Arbitration Act 1996 and not under schedule II of the 1986 Act with all its requirements and time-limits.

Among the changes made by the 2006 Order are the following:

(a) there is now no Lord Chancellor's Panel of Agricultural Arbitrators. Now the arbitrator is to be appointed by agreement or by the President of the RICS if a name cannot be mutually agreed

(b) amends section 84 of the 1986 Act. The effect is that now all arbitration under the 1986 Act will now be under the Arbitration Act 1996

(c) there are now no statutory grounds to insist on a Welsh speaking arbitrator in Welsh disputes.

24.4 Appointment procedure

When an issue arises, eg rental revision, failure to agree an outgoing tenant's claim, or a landlord's counterclaim for dilapidations (in the latter two cases where eight months have elapsed since the termination of the tenancy), it is usual to agree the name of a person to act as an arbitrator. It is normal for one party to submit to the other three names of possible arbitrators and if any are acceptable that person is chosen and appointed. However, in practice, it is sometimes even difficult to achieve agreement on any one name. In such cases, an application, on the prescribed form, (see example) enclosing the prescribed fee (at the time of writing £115) is made to the President of the RICS.

24.5 The President will make an appointment and from that date the person appointed takes on his role as arbitrator. In due course when this appointment is made the parties are notified by the President.

24.6 Arbitrator's procedure

On receiving his appointment the arbitrator will write to the agents appointed by the parties (or if no agents' names are given, to the parties direct) notifying them of his appointment, the date of his appointment and requesting the delivery to him, in duplicate, within a specified date of his appointment, of statements of their respective cases with all necessary particulars.

24.7 The statement of case

Rental arbitrations

Schedule 2 of the Agricultural Holdings Act 1986 sets out the matters which an arbitrator must take into account in assessing the rent payable in respect of a holding.

The statement of case should therefore be drafted in line with the matters which the arbitrator has to take into account. Prior to the passing of the Agricultural Holdings Act 1986 the tendency among many valuers was to deliver rather brief and sketchy statements of case. This is an unwise course to adopt and it is advisable that a great deal of detail should be included in statements of case.

The following should be set out in the statement of case.

1. Introduction

This should:

(a) give the name of the owner, the name of the tenant, the address of the holding, describe briefly the holding, the type it is and give its area

(b) state the date the tenancy commenced, the rent then passing, subsequent revisions and also the current rent, and when this was fixed

(c) if additional land was included in the tenancy give the date of such inclusion and also the additional rent payable

(d) state when the notice requiring arbitration was served and the date (if applicable) when the application was made to the President of the RICS for the appointment of arbitrator. It should also state the date on which the arbitrator, giving his name, was appointed.

2. Terms of the tenancy

Detailed reference should be made to the tenancy agreement, the term date, dates when the rent is payable, if in advance or arrears, repairing liabilities, responsibility for insurance and any other matters in the tenancy agreement that are considered to be relevant, especially any unusual covenants.

A plan should be provided.

3. Character and situation of the holding

The situation of the holding should be given and also a description of the holding provided with details of the residence, cottages (how occupied) the buildings and the land. Reference should be made to the topography, the type of soil including references to soil types (obtained from the soil survey), quality (grading of soil) and its crop-bearing capacity. Also, availability of water supplies, altitude of the land, accessibility, etc.

4. Productive capacity

Details should be given of the farming system practised, including particulars of stock kept, crops grown and if so considered, give the opinion that the system practised is the best for the holding. If production cannot be increased this should be stated.

5. Relating earning capacity

Here reference to a budget for the next year should be made to illustrate the earning capacity that can be expected to follow the productive capacity. It may not be convenient to provide a budget at this stage and if so, it should be stated that this will be provided at the hearing. This budget should be in great detail and should show the profits expected to be earned. Ideally the budget should be for the next three years, but in practice an arbitrator will be content with the first year's budget, well knowing the volatility of supply and demand for agricultural products and resultant ups and downs of prices and consequently the profits that can be expected to be earned.

Note: it is advisable, to save the arbitrator's time at the hearing, to have prior agreement of the budget with the agent's opposite number, but in practice it is rare for a budget to be totally agreed.

6. Level of rents of comparable lettings

If good information, supportive of the case is available, reference to specific comparisons to be cited should be made. (In practice, it is difficult except in the case when the holding is on an estate where the landlord has the information, to provide comparisons.) Other tenants rarely like their business to be bandied about and are reluctant to divulge the necessary information or to give permission for their farms to be inspected. If comparisons are to be given, the name and address of the comparable farm(s) should be provided including all the terms of the tenancy, rent passing and when last revised, tenant's improvements, etc, and the tenancy agreement should be provided. An opportunity should be given for the opposing party to inspect the comparison prior to the hearing and also to inspect the tenancy agreement.

If no comparison is to be given, reference here can be made to Rent Reports, or other evidence available cited, in a general way, in order to illustrate rental trends.

7. Tenant's improvements

Under this heading a detailed list of tenant's improvements (executed with or without landlord's permission) should be given, including dates carried out. It is unnecessary to give costs (which historically may be low). The rental value equivalent of the tenant's improvements can be given (but it is not necessary to provide details of the calculation at this stage), or in adopting the black patch approach ask the arbitrator to disregard these improvements.

If a claim is being made by the landlord for additional rent in respect of landlord's improvements (discounting grant aid received) this should be stated, giving full details of the improvements, and, in this case the amount of additional rent sought.

8. Rental valuation

The rental considered to be properly payable and sought to be awarded should be stated, but at this stage it is not necessary to provide details of how the valuation is calculated.

9. General

Mention should be made of the following, if applicable, and dependent on whether it is the landlord's or tenant's statement of case:

- the demand for holdings and scarcity of holdings to let
- any particular disadvantages, in a general way the holding suffers from and what abnormal difficulties or problems may be encountered in farming the holding
- any special features and advantages the holding possesses
- any lack of fixed equipment
- any other matter of relevance to the issue of the rent which is properly payable.

10. Costs

A statement should be made that the arbitrator will be addressed on costs.

24.8 The hearing

When both statements of case are received, the arbitrator will exchange them. If it is a complicated case he may fix a preliminary hearing when he may inform the parties that he requires discovery of certain information bearing on the issue and a certain time-limit will be given for such information and documents to be produced. He may also ask the permission of the parties to engage a legal advisor (he cannot engage a legal advisor and charge his costs without the permission of the parties — but this permission is rarely withheld). The venue for the hearing is often a room at the farmhouse, a local inn or village hall. The arbitrator will arrange the seating, having his legal advisor (if any) at the top table, and the parties facing each other at a table in front of the arbitrator. A place, normally facing the arbitrator will be reserved for witnesses to give evidence.

In recent years, in order to save time and expense it has become practice for the parties to exchange witness statements and budgets before the hearing. This is something that should be encouraged. The arbitrator is really a judge of the matters at issue and decides the issues on the evidence before him. It is important that he should remember this and control the proceedings in a dignified, formal and courteous manner but at all times he should be in complete control. He should not allow smoking. Strangers who have nothing to do with the proceedings should not be allowed to attend as the proceedings are private. The arbitrator should act in a totally impartial manner and should not show favour to anyone. An offer of coffee or luncheon made (it frequently is) should be refused as the receipt of any form of hospitality from one party may be strongly objected to by the other.

The hearing procedure more or less follows court actions and is as follows:

1. the claimant (or rather his advocate) opens his case and proceeds through this
2. he calls his witnesses who are put on oath by the arbitrator
3. witnesses are cross-examined
4. witnesses may be re-examined

5. the respondent (or rather his advocate) opens his case and proceeds through this
6. the respondent witnesses are called to give evidence on oath
7. witnesses are cross-examined
8. witnesses re-examined
9. the respondent sums up his case
10. the claimant sums up his case
11. costs — the claimant or his advocate address the arbitrator on costs and this is followed by the respondent making an address on costs.

The advocate/valuer usually produces a detailed, written witness statement or proof of evidence and budget for his client, and all witnesses are also advised to have written proofs of evidence. The witness, after being sworn, reads his proof of evidence. In this way important points and evidence are not missed out, and time is saved. Copies of the witness' proof of evidence are supplied to the arbitrator and the other party after the witness is sworn. However, it is most advisable in order to save costs that such proofs of evidence or witness statements are exchanged before the hearing, as time will be saved in that the witness does not have to read this statement and will be merely examined on salient points and then cross-examined.

The arbitrator will be taking notes of the evidence given, and by receiving proofs of evidence and documents during the proceedings. Sometimes the proceedings are recorded or a stenographer is present. If a valuer is acting as both an advocate and giving evidence, the arbitrator will request him to take the oath or affirmation.

24.9 The inspection

Some arbitrators make it their practice to make a brief inspection, alone, before the hearing is held, especially in rental arbitrations in order to acquaint themselves with the property concerned. This is much to be commended as the arbitrator will have the advantage of knowing his subject much better than where he is a total stranger to the property.

Inspections are normally made at the conclusion of a hearing or, if there is insufficient time, on another day. Most arbitrators prefer to make inspections unaccompanied unless it is essential that the parties or their valuers should be present to point out matters which were the subject of an issue at the hearing. If one party is present then the other or his representative should also be there to avoid any chance of undue representations being made, in his absence, to the arbitrator.

24.10 The award

There is now no presumed form of the award. It is up to the arbitrator to decide an appropriate form of award. If the parties cannot agree the form it is to:

- be in writing and signed by the arbitrator
- contain the reasons for the award (unless the parties agree to dispense with this)
- state the seat of the arbitration and the date when it was made.

There is now no statutory time-limit for making the award nor payment of monies due under it.

24.10.1 Special case

An arbitrator may at any stage of the proceedings and shall if so directed by the county court, state a special case for the opinion of the county court on a question of law arising in the arbitration or as to his jurisdiction.

24.11 Costs

The costs of, and incidental to, the arbitration and award shall be at the discretion of the arbitrator. Such costs can on the application of either party, be taxable in the county court.

The arbitrator shall in awarding costs take into consideration the reasonableness or unreasonableness of the claim, of either party, and any unreasonable demand for particulars or refusal to supply such and generally the circumstances of the case.

It is now widely held that costs should follow the event, ie the loser should pay. This is in sharp contrast with the practice adopted extensively in the past by many arbitrators, of dividing costs. However a Calderbank offer (*Calderbank* v *Calderbank* [1975] 3 All ER 333) may have a major bearing on the award of costs that have arisen after the offer is made. A Calderbank offer can be done by submitting to the opposite party a written offer in settlement which is made 'Without prejudice except as to costs'. This offer is made well in advance of the hearing and, if not accepted, the arbitrator at the hearing is handed a sealed envelope which is marked 'to be opened after the interim award, in the presence of the parties or their agents, but before determination of costs'. If the arbitrator's award is the same or less than the Calderbank offer, then he will normally award to the winning party the costs which have arisen since the offer was made to the other party. The arbitrator in issuing the interim award will give no direction on costs. Subsequently if no sealed offer is tendered at the hearing, the parties will send him copies of any Calderbank offers that they have, prior to the hearing, passed between themselves and the arbitrator then will publish his final award which then deals with the matter of costs.

24.12 Procedural points at the hearing

1. The arbitrator opens the proceedings by issuing an appearance list for completion by those attending.
2. He states who he is and that he is acting as an arbitrator and then should state who has appointed him, whether the parties or the President of the RICS, the date of his appointment and also state the matter which is referred to his decision.
3. He then asks the advocate/valuers for their formal appointments to act. In cases where the President of the RICS has made the appointment, formal appointments are not necessary.
4. He reminds the parties of the normal procedure that is adopted in such hearings.
5. When witnesses are called, he administers the oath or affirmation.
6. He should examine all documents submitted in evidence and see that the opposite party has a copy of the same. In particular, where documents should be stamped, he should not receive them in evidence unless they are stamped or the cost of stamping together with the penalty for late stamping is deposited there and then to the arbitrator.
7. If an adjournment is sought this should be granted unless it is clearly a time-wasting and stalling device. However, the view of the opposing party should always be sought before the arbitrator decides whether or not to grant an adjournment.
8. He should fix a time for the inspection.

24.13 Arbitrations other than rental

There are many other matters regarding the subject of agricultural arbitrations, other than rental, and these include:

- succession on retirement — the rental payable and the terms of the tenancy
- succession on death — the rent payable and the terms of the tenancy
- terms of a new written tenancy agreement where one did not previously exist
- reduction of rent following the landlord obtaining possession of part of the holding
- damage by game — compensation payable
- settlement of claims by outgoing tenants for tenant right, improvements and fixtures
- settlement of the landlord's claim for dilapidations and deterioration.

24.14 Statement of case

Again, as in the case of rental arbitrations there is no official form of statement of case prescribed, but broadly the statement should, apart from giving details of the tenancy, refer to the tenancy agreement, give the names of the landlord and tenant, the termination date of the tenancy (where relevant), show the amount of the claim in detail (where applicable), give the section(s) of the Act under which the claim is made and where applicable, eg in the case of a dilapidation or tenant right claim, give fully itemised particulars of the claim showing each separate item of claim and the amount of claim against each item. There is now no statutory time-limit for the submissions of statements.

Example rental arbitration

Old Hendre, Ross-on-Wye is a 403.01 acre old red sandstone arable farm with Georgian farmhouse, modern Woolaway 3 bed bungalow, one other cottage (let) and an extensive set of buildings including 500 ton grain store erected by the tenant six years ago. The soil is easy working, Grade 11 (60%), Grade III (40%) on the Land Classification Map. The farm has extensive frontage to the River Wye (which floods 80 acres) and the tenant has obtained an irrigation licence.

It is owned by the Aral Estate Co and is let to Mr J Yeomans on an annual Candlemas (2 February) tenancy which commenced 11 years ago. Mr Yeomans obtained the tenancy by open tender and at the time paid a rent of £28,250 per annum on a full repairing and insuring agreement. During the negotiations he contended that when he obtained the tenancy of the holding he added an overbid of 25% of the rent payable as key money, in order to secure the tenancy. The rent has remained unaltered since. The landlord served on 28 January 2007 a notice under section 12 of the Agricultural Holdings Act 1986 requiring arbitration on the rental. The landlord's agents have been seeking a revised rental of £35,250 per annum. The tenant has, reluctantly, offered an increase to £30,000 which has been refused. The parties have failed to agree an arbitrator and an application was made by the landlord, on 3 January 2008 to the President of the RICS for an arbitrator to be appointed. The President appointed Mr IW Wright FRICS, FAAV as arbitrator on 28 January 2008.

The farming system practised is growing 100 acres winter wheat, 80 acres winter barley, 40 acres spring barley, 35 acres potatoes, 65 acres sugar beet and the remaining 80 acres is pasture keeping 400 breeding ewes. 100 beef bulls are produced annually. 500 root tegs are produced having been purchased in November of each year as stores. Mr Yeomans is 46 and keeps two men. Mr Yeomans also farms 75 acres which adjoins the farm and which he purchased from the landlords five years ago. He also takes on a seasonal grazing licence — 15 acres of pasture nearby.

Tenancy agreement
Dated 4 February 1987. A normal tenancy agreement except that the tenant carries out all repairs and pays landlord's fire insurance premium.

Tenant's improvements
Erection of 500 tonne grain store. Tile drained 20 acres. Installed central heating in farmhouse. Installed 3 phase electricity supply.

Arbitration Act 1996
Agricultural Holdings Act 1986

Application for the appointment of an arbitrator for use in s12 rent cases

To the President, The Royal Institution of Chartered Surveyors, (Arbitrations Section), Surveyor Court, Westwood Way, Coventry CV4 8JE

A.   The landlord and the tenant having failed to agree as to the person to act as arbitrator and there being no provision in any agreement between them relating to the appointment of such arbitrator. We West & Stone. 6 High Road, Ross-on-Wye. Herefordshire HR9 6EO, hereby apply to the President of the Royal Institution of Chartered Surveyors for the appointment of an arbitrator as to the rent to be paid for the holding referred to below as from the next termination date following the date of the demand for arbitration served by the landlord on 28 January 2007 on his tenant.

B.   We, West & Stone enclose a cheque for £115 made out to The Royal Institution of Chartered Surveyors.**

C.   We, West & Stone understand that unless the application, accompanied by the fee, is received at the address given above before the next termination date following the date of the demand for arbitration, it will be invalid and the appointment of the arbitrator will not be made.

West and Stone . . . .   3 January 2008 . . . .
Signature.   Date . . . .

State whether landlord or tenant or duly authorised agent of either

Agents for the Landlord

. . . .

PARTICULARS REQUIRED	REPLIES
1. Name and address of the agricultural holding (as defined in section 1 of the Agricultural Holdings Act 1986).	Holding(s): Old Hendre Parish: Ross-on-Wye, County: Hereford
2. Name and address of landlord	Aral Estate Co, c/o Agents West & Stone
3. Name and address of landlord's agent (please quote reference)	West & Stone, 6 High Road, Ross-on-Wye, Herefordshire
4. Name and address of tenant	Mr J Yeomans, Old Hendre, Ross-on-Wye, Herefordshire
5. Name and address of tenant's agent (please quote reference)	Mr JT Thomas FRICS FAAV, FSVA, The Hollies, Ross-on-Wye
6. Approximate area of holding	403 acres
7. Description of holding (for example, mixed, arable, dairying, market garden)	Mixed arable and stock farm
8. Has a demand in writing for an arbitration as to the rent to be paid for the holding been made by one party to the other? If so, state the date of such demand and whether it was made by the landlord or the tenant.	Yes on 28 January 2007 made by the landlord

PARTICULARS REQUIRED	REPLIES
9. State the next termination date following the date of the demand.	2 February 2008
10.(a) On what date did the tenancy commence?	2 February 1997
(b) On what date did any previous increase or reduction of the rent take place?	No alteration in rent since 2 February 2007
(c) On what date took effect any previous direction of an arbitrator that the rent should continue unchanged?	N/A

**VAT is not payable and the fee is non-returnable. If more than one holding is to be referred to arbitration, a fee is payable (and a separate form must be submitted) in respect of each holding.

IA Wright Esq FRICS, FAAV
Chartered Surveyor,
2184 High Town,
Hereford   BY RECORDED DELIVERY

To: Messrs West & Stone, 6 High Road, Ross-on-Wye.
         Mr J Thomas, Chartered Surveyor, The Hollies, Ross-on-Wye.

Dear Sir
AGRICULTURAL HOLDINGS ACT 1986
HOLDING: OLD HENDRE FARM, ROSS-ON-WYE, HEREFORDSHIRE
LANDLORD: ARAL ESTATE CO
TENANT: MR T YEOMANS

I write to confirm that I have been appointed by the President of the Royal Institution of Chartered Surveyors to act as arbitrator in this case to determine the rent properly payable for the above holding as from the next termination date following the demand for arbitration made by the landlord to the tenant. My appointment is dated 28 January 2008. The following is the procedure I intend adopting unless any party objects, in which case I will call a preliminary hearing of the parties.

1.   **Statements of case**
Statements of case must be submitted to me in duplicate by each party within 35 days of the date of my appointment. On receipt of both, copies of each will be sent to the other party.

2. Date of hearing and place

It would be helpful if you could suggest a convenient date, about 6/7 weeks hence, for the hearing. Also a suggested venue for the hearing.

3. Representation

Please let me know if the parties will be legally represented at the hearing either by counsel or by a solicitor. Please also let me know how many witnesses, if any, each party will be likely to call.

4. Costs

My costs in this matter will be £120 for opening the file and then at the rate of £150 per hour including time spent considering representations, travelling, etc, plus all disbursements and out-of-pocket expenses and VAT. I would ask both parties to confirm agreement to this before the hearing. Unless I hear to the contrary I shall assume that this level of costs is acceptable.

5. Generally

May I remind you that all communications with me should be in writing, and that both parties should ensure that any letters written to me should be copied and simultaneously despatched to the other party.

I would also remind the parties that no inadmissible evidence should be produced and no 'Without Prejudice' discussions, communications or negotiations should be referred to in any way if they have not resulted in any agreement.

I look forward to hearing from you on this matter as soon as possible.

Yours faithfully

IA Wright FRICS, FAAV, Arbitrator

AGRICULTURAL HOLDINGS ACT 1986
IN THE MATTER OF ARBITRATION BETWEEN:

ARAL ESTATE CO (Landlord) and Mr J YEOMANS (Tenant)
Concerning the rent to be properly payable in respect
of the Holding known as:
OLD HENDRE, ROSS-ON-WYE
In the County of HEREFORD

STATEMENT OF CASE for THE ARBITRATOR with particulars
on behalf of the Landlord

1. Introduction

Old Hendre Farm, situated in the Wye Valley, 4 miles south of the market town of Ross-on-Wye is a 403 acre arable and stock farm owned by the Aral Estate Co and is tenanted by Mr J Yeomans.

2. Tenancy

The tenancy commenced on 2 February 1997 and the rent payable at that time was £28,250 per annum. There has been no subsequent rental revision in the intervening period and the rental passing is still this sum. A notice requiring arbitration was served under recorded delivery by the landlord's agents on 28 January 2007.

Subsequently, negotiations have taken place firstly between the landlord's agents and the tenant direct, and as matters could not be agreed an application was made on 3 January 2008 to the President of the Royal Institution of Chartered Surveyors to appoint an arbitrator and Mr IA Wright, FRICS FAAV was appointed on 28 January 2008 by the President to determine the issue.

3. Terms of the tenancy

The holding is held under a written tenancy agreement dated 4 February 1997 (a copy of which is annexed). This is a normal tenancy agreement (which will be provided in evidence) except that the tenant has undertaken to be responsible for all repairs and the payment of the landlord's fire insurance premium.

4. Character and situation of the holding

Old Hendre is a very good quality arable and mixed farm well situated on a council road 4 miles south of Ross. It is capable of growing most kinds of cash crops. It has a good quality 5 bedroom Georgian residence, extensive modern and traditional buildings and good quality land with soil being derived from the old red sandstone formation being Grade II and III soil on the Land Classification map. The land is either level or gently undulating and a substantial area has frontage to the River Wye.

The holding has two cottages, one occupied by the tenant's workman and the other sub-let by him (with landlord's permission).

5. Productive capacity and related earning capacity

The holding is primarily used for arable production but a ewe flock and a bull beef enterprise are run from the farm. Wheat, barley, potatoes and sugar beet are grown.

A budget illustrating the productive and thus related earning capacity of the holding will be prepared and submitted in evidence to the arbitrator.

6. Levels of rents and comparable lettings

Evidence will be given of a comparable letting on the Aral Estate and the general level of rents now payable not only on this Estate but also regionally and nationally.

7. Tenant's improvements

The tenant has carried out the following improvements:

7.1 Erection of a 500 tonne grain store
7.2 Drained OS 560, 20.12 acres
7.3 Installed 3 phase electricity

7.4 Installed central heating in the farmhouse. It is, however, observed that while 7.1–7.3 were beneficial to the holding, 7.4 was for the tenant's personal comfort

7.5 Carried out some improvements to the house.

8. General
Reference will be made to:

8.1 The general scarcity of holdings to let and the extensive demand for tenancies

8.2 The amount of the SFP the tenant receives which in 2006 totalled £35,061

8.3 The trends and other factors affecting farm rents

8.4 All other matters considered relevant to the rent of the holding.

9. Rental valuation
Taking into account all the above matters and having regard to current market conditions and the terms of the tenancy, we consider that the rental properly payable for the holding under the provisions of section12 and schedule 2 of the Agricultural Holdings Act 1986 and taking into account all relevant factors, is in the sum of £37,500 (thirty-seven thousand, five hundred pounds) per annum.

The tenant is also, in addition to Old Hendre, farming 75 acres of land immediately adjacent and which he purchased five years ago from the landlord. The homestead and buildings serve this land. In addition he has for several years taken a seasonal grazing licence on 15 acres of pasture belonging to the landlord. It is therefore considered that an addition for marriage value (*Childers* v *Ankers* CA1995) is payable and this is included in the rental sought.

10. Plan
A plan of the holding will be provided at the hearing.

11. Evidence
The right is reserved to enlarge on the above at the hearing and witnesses will be called as appropriate to give evidence.

12. Costs
The right is reserved to address the arbitrator costs.

Dated this 28 day of February 2008

Signed . . . .
JR Andrews FRICS, FAAV
for West & Stone, Agents for the Landlord
Chartered Surveyors and Land Agents
6 High Road, Ross-on-Wye,
Herefordshire HR9 6EQ   E&OE

AGRICULTURAL HOLDINGS ACT 1986
IN THE MATTER OF ARBITRATION BETWEEN:

ARAL ESTATE CO (Landlord) and Mr J YEOMANS (Tenant)
Concerning the rent of the property payable in respect
of the Holding known as:
OLD HENDRE, ROSS-ON-WYE in the county of HEREFORD

STATEMENT OF CASE for THE ARBITRATOR with particulars
on behalf of the Tenant

1.   Introduction
Old Hendre Farm, is a 403 acre mixed arable and stock farm situated 4 miles from Ross-on-Wye, half a mile off the B1234 road and is in the Wye Valley. It is tenanted by Mr J Yeomans from the landlords, the Aral Estate Co.

2.   Tenancy
The annual Candlemas agricultural tenancy commenced on 2 February 1997 at a rental of £28,250 per annum. A notice requiring arbitration on the rental was served by the landlord's agents on 28 January 2007.

Subsequent negotiations have failed to resolve the matter and the President of the Royal Institution of Chartered Surveyors appointed Mr IA Wright FRICS, FAAV as arbitrator to decide the issue of 28 January 2008.

3.   Terms of tenancy
The holding is held under the terms of a written tenancy agreement dated 4 February 1997. The tenancy commenced on 2 February 1997 and is an annual tenancy with rental paid half yearly, in arrear on 2 August and 2 February. It is a standard tenancy agreement excepting that the tenant has the abnormal burden of being responsible for all repairs and paying the costs of fire insurance of the fixed equipment.

4.   Character and situation of the holding
The holding is primarily an arable farm which is either level or part undulating and a large part of the land is subject to flooding by the River Wye. It has a substantial residence and adequate buildings, part modern, part traditional. There are also included a pair of cottages.

5.   Productive capacity and related earning capacity
The arable enterprise comprises: growing wheat, barley, potatoes, sugar beet, and approx. 80 acres of grassland (liable to flooding) maintains the 400 breeding ewes kept. About 500 root tegs are fattened in winter and there is also a bull beef unit with an output of about 100 head annually.

A budget illustrating the productive capacity and thus the related earning capacity of the holding will be produced at the hearing. Also the past three years' farm accounts will be provided.

6. General level of rents

The latest available rental statistics will be produced and referred to, showing the virtual standstill of rentals of holdings of this nature.

7. Tenant's improvements

The tenant has carried out the following improvements at his sole expense:

7.1 Erected steel and concrete 500 tonne grain store
7.2 Installed 3 phase electricity supply
7.3 Tile drained 20 acres — OS No 560
7.4 Installed oil fired central heating in the farmhouse — 15 radiators
7.5 Installed a new kitchen and bathroom and generally improved the dwelling.

8. General

Reference will be made at the hearing to the following relevant matters affecting the rental payable for the holding:

8.1 The constantly diminishing level of profits experienced by farmers and the tenant, in particular, in recent years, particularly in 2007
8.2 The annual reduction of Single Farm Payments and further reductions forecast
8.3 Constantly increasing fixed and variable costs incurred
8.4 The increasing pressure mounted on agriculture by the conservation and green element of the community and general 'farmer bashing' which ultimately affects demand and increases costs. The extensive use by the public of the Wye Valley footpath along this farm results in losses and nuisance from trespass.
8.5 When the farm was originally let, the rental paid, by tender, was at least 25% above the level of rents, paid for similar holdings. A substantial 'Key' element was included in the tender.
8.6 Any other relevant matters.

9. Rental valuation

Having taken into account all the above and other relevant matters it is considered that the rent properly payable for the holding, having regard to the onerous terms of the tenancy is £30,000 per annum.

10. Evidence

The right is reserved to fully amplify the above statement at the hearing and witnesses will be called in evidence.

11. Costs

The arbitrator will be addressed on costs.

Dated this 25 day of February 1998

Signed . . . .
JT Thomas FRICS, FAAV
Chartered Surveyor, Agent for the Tenant
The Hollies, Ross-on-Wye
Herefordshire   E&OE

AGRICULTURAL HOLDINGS ACT 1986
INTERIM AWARD

Arbitrator:	Ian Arthur Wright FRICS, FAAV
	2184 High Town, Hereford
Date of Appointment:	28 January 2008
Landlord:	Aral Estate Co, c/o West & Stone,
	6 High Road, Ross-on-Wye
Tenant:	Mr J Yeomans, Old Hendre, Ross-on-Wye
Rent payable prior to arbitration:	£28,250 per annum

AWARD OF THE ARBITRATOR

The claims or questions set out in the Schedule to this award have been referred to arbitration and, having considered the evidence and the submissions of the parties, I, the arbitrator, award as follows:

1. As from 2 February 2008 the next termination date following the date of the demand for arbitration the rent perviously payable is increased to £32,500 per annum.
2. The cost of and incidental to the arbitration and the award shall be dealt with in my final award when I have been given details of Calderbank offers to settle, made by the parties to each other, 'save as to costs' prior to the hearing when I will provide, as requested, reasons for my award.

Signed by the arbitrator IA Wright in the presence of:
J Anderson,
2184 High Town, Hereford

Date: 30 April 2008

AGRICULTURAL HOLDINGS ACT 1986
FINAL AWARD

OLD HENDRE ROSS-ON-WYE HEREFORDSHIRE

Arbitrator:	Ian Arthur Wright, FRICS, FAAV,
	2184 High Town, Hereford.

Date of Appointment: 28 January 1998
Time for making award
extended to: 30 May 1998
Present Landlord: Aral Estate Co, c/o West & Stone,
                                        6 High Road, Ross-on-Wye.
Present Tenant: Mr J Yeomans, Old Hendre, Ross-on-Wye.
Rent payable prior to arbitration: £20,610 per annum.

AWARD OF THE ARBITRATOR

The claims and questions set out in the Schedule to this Award have been referred to arbitration, and having considered the evidence and submissions of the parties, I, the arbitrator, award as follows:

As from 2nd February 1998 (the next day on which the tenancy could have been brought to an end by notice to quit at the date of the notice demanding arbitration under Section 12 of the Act) the rent previously payable is increased to £27,000 per annum being the rent properly payable in respect of the holding at the date of the reference to arbitration.

COSTS
AWARD AND DIRECT

1. The landlord shall pay my costs of this reference amounting to £6,020 plus VAT of £1,053.50.
2. As respects the cost of and incidental to this arbitration, the landlord must pay the costs of the tenant, which failing agreement, shall be taxed in the County Court according to Scale II as prescribed by the County Court.
3. The landlord shall bear his own costs of this reference.
4. Any costs payable by one part to the other party under or by virtue of this award shall be so paid after the delivery of this Award.

Signed by the Arbitrator I A Wright in the presence of:
J Goodhew (Miss)
8 Fair Oaks, Ross. Secretary.

Dated this 16 May 1998

This Award was delivered to the landlord's agent, Messrs West and Son, on 20 May 2008.

THE SCHEDULE

The rent payable for the farm from 2 February 2008 being the next termination date following the date of the demand for arbitration on 28 January 2007.

Rent awarded £32,500 per annum. This interim award was delivered to the landlord's agents Messrs West & Stone on the 6th day of May 2008.

Statement of reasons
The reasons for this final award are attached.

STATEMENT OF REASONS FOR THE AWARD AND COSTS

1. Why the arbitration arose
The arbitration arose as a result of the landlords, the Aral Estate Co, and the tenant, Mr J Yeomans, failing to agree on a revised rental payable for Old Hendre, Ross-on-Wye in the County of Hereford with effect from 2 February 2008, following the service of a section 12 notice requiring arbitration, on the question of the rental, served by the landlord on the tenant on 28 January 2007.

2. Agreement of facts
I find that the following facts were agreed between the parties, either in the statement of case submitted and/or during the course of the hearing held on 6 March 2008.

2.1 The tenant held Old Hendre under the terms of a written tenancy agreement dated 4 February 1997 and the tenancy commenced on 2 February 1997.
2.2 The rental payable, with effect from 2 February 1997 is £28,250 per annum, and this has not been varied since.
2.3 The tenant is responsible for all repairs to the holding and also for payment of the landlord's insurance premium.
2.4 Apart from the tenant's responsibility for all repairs and insurance, there are no particular unusual clauses in the tenancy agreement except that in clause 38 the tenant is allowed enter into contracts (ie to partnership farm) with producers of specialist crops up to 60 acres in any one year, on the terms set out in this clause.

3. Valuation principles to be adopted in considering the rent properly payable for the holding
In determining the rent properly payable, schedule 2.1 of the Agricultural Holdings Act 1986 directs that various matters should be taken into account. I set out the submissions made and evidence given and my findings in relation to the various matters to be taken into account as follows.

3.1 Terms of the tenancy
Evidence
3.1.1 The holding was originally let by tender under the terms of the written agreement dated 4 February 1994.
3.1.2 The area of the farm is still the same as when originally let, 403.01 acres.
3.1.3 The repairing and insuring obligations are a departure from the standard in that the tenant is responsible for all repairs and also responsible for reimbursing the landlord for the insurance of the

fixed equipment. The insurance policy is a comprehensive cover policy. Evidence was given by the tenant that the average annual cost of all repairs to him since the commencement of the tenancy was £6,210 and last year landlord's insurance cost £3,756.

3.1.4 The term date is Candlemas (2 February) in each year, with rental payable quarterly in arrears on 2nd day of August and February in each year.

3.1.5 The tenant is allowed under clause 38 of the tenancy agreement to enter into contracts with producers of specialist crops of up to 60 acres in any one year. On cross-examination, the tenant gave evidence that in 2007 he had, in partnership with a neighbour, grown 35 acres of potatoes. This had produced him a net return, less the rent paid, of £300 per acre.

Findings

I find that the following are unusual terms on normal and customary lettings of similar proportions.

(a) The tenant is responsible for all repairs. Usual repairing liabilities are in accordance with the Agricultural (Maintenance, Repair and Insurance of Fixed Equipment) Regulations 1973 (SI 1973 No 1473) (as amended) where landlord and tenant share the repairing responsibility as laid down in the SI.

(b) The tenant is responsible for the landlord's insurance costs. Landlords, under SI 1473, are normally responsible for insurance.

(c) The tenant is allowed to enter into contracts with specialist growers up to 60 acres in any year.

I find that (a) and (b) are additional burdens above the ordinary on the tenant, but (c) is a possible advantage to him.

3.2 **Character and situation of the holding**

Evidence and findings

Taking into account evidence given, documents put in evidence and my subsequent inspection of the holding, my findings are:

3.2.1 Old Hendre is a very good quality mixed arable and stock farm, approached by a rather narrow, unsurfaced and rather pot-holed drive, 800m long from the B1234 Ross-Lydbrook road, about 4 miles south of Ross-on-Wye. Apart from the access drive, the approach is fair.

3.2.2 The farm is, with the exception of OS Nos 5568, 30.45 acres, in a ring fence. It is bounded on the east by the B1234 road and on the west by the River Wye.

3.2.3 The farmhouse is a very substantially built and attractive brick and slate dwelling, standing on its own with good views. It is a spacious and well designed dwelling. It is surrounded by extensive well maintained grounds.

The tenant has carried out improvements to the house as detailed below.

3.2.4   The farm buildings are generally extensive in nature and I find they are more than adequate for a farm of 403.01 acres. The landlord's buildings are generally of the traditional type, but are all useable. The 500 tonne grain store erected by the tenant is an excellent building which is needed on this holding.

3.2.5   The land is good quality land, generally easy working and the soil is derived from the old red sandstone formation. 80 acres or thereabouts are on the flood plain of the Wye and evidence was given that in some winters up to seven high floods have been experienced. The tenant gave evidence that he has had considerable expense in ditch maintenance on the flood land. Approximately 60% of the farm is Grade II and 40% Grade III on the land classification map.

Further evidence was given by the tenant that approximately 60 acres on the eastern side of the farm — the higher land — dries out quickly in dry times and crops suffer accordingly.

I find that this is generally good land, very well farmed, in excellent heart but has constraints due to flooding and drying out. There are increased costs above normal, incurred in cleaning out extensive ditches on the flood plain land.

3.2.6   The property is served with mains water and electricity with either a natural or piped water supply to every field.

3.2.7   The western part of the holding is crossed by a 400kv overhead electricity line for distance of 800m. This restricts irrigation which is available from the River Wye for most of the holding. I find the overhead line restricts irrigation for an area of approx 28 acres.

3.3   **The productive capacity of the holding**

Evidence

The evidence of the tenant was that the holding was utilised on a rotational arable and livestock system which he considered was the most suitable to obtain maximum production.

Evidence was further given that the tenant was farming, in addition to Old Hendre of 403.01 acres, further land as follows:

Land he owned adjoining	75 acres
Grazing licence land adjoining	15 acres

This added to Old Hendre makes a total of 493 acres or thereabouts. Currently the tenant grows (or plans to grow in 2008) arable crops including wheat and barley, as well as farming 35 acres of potatoes on contract. In addition, he keeps 400 breeding ewes, 500 root tegs and has an annual output of 100 beef bulls sold on contract.

The landlord's expert agricultural witness Mr FA Knoall BSc (Agric), in his evidence maintained that the best farming system to be utilised at the holding was an all arable system growing sugar beet, winter wheat/winter barley /spring beans in rotation. The remaining 40 acres is proposed to be in permanent pasture or long-term leys used for grazing and finishing 80 head of stores, summer grazing and winter finished on silage and barley in the buildings. No sheep are to be kept.

This system, in the witnesses' view is the best system to adopt to make the best possible use of the highly productive soil the holding possesses.

The proposed system will, in the view of the witness, reduce the labour requirement of the farm to the tenant and one man. Casual labour and contractors are to be used during planting and harvest.

The effect of the landlord's proposal, in Mr Knoall's view, is that one current full-time craftsman employee would be redundant and that the cottage he occupies would not be needed to house him and can be surrendered to the landlord.

This, all in all, in the view of the landlords, would increase gross margins and reduce costs thus showing a higher division of the profits for sharing between the parties.

The tenant's expert witness, Mr JR Littleman, FRICS, FAAV on the other hand supported the tenant's existing system in growing 145 acres winter wheat, 80 acres winter barley, 40 acres spring barley (for malting), 35 acres potatoes, and using the remaining 80 acres as pasture for hay and silage, keeping 400 ewes, finishing 100 beef bulls and wintering 500 root tegs.

This system involves utilising two workmen (one of whom resides in a cottage on the farm) and the tenant. The witness gave evidence that most of the additional land proposed to be ploughed was totally unsuitable for arable production as some of the fields are wet and some on a slope with shallow soils, and in his view should not be ploughed.

The tenant in his evidence, disputed that it would be practical to plough up the fields the landlord's expert considered should be ploughed as the land was totally unsuitable for arable cropping due to its wet nature, undulation, steepness and shallow soils in parts. He also maintained he could not change his farming system at short notice for what might be a short-term gain. He maintained that his present farming system is what a prudent tenant would practice and also gave evidence that in the last eight years he had won, in the local agricultural societies' competitions, a total of 40 crop prizes (30 1st, five 2nd and five 3rd) including the best farmed farm over 300 acres within a radius of 15 miles of Ross on three occasions.

He further gave evidence that he needed to keep a substantial ewe flock on this farm, to maintain fertility on much of the light soil it possesses.

Also, that a prudent tenant would keep sheep since he needed cash flow in spring and early summer when he had little else to sell.

Findings

I find that the current enterprises carried out at this farm are the most suitable for this holding. I find the tenant is a prudent and able farmer who, I consider, is using this holding to its maximum potential. I reject the landlord's submission that the tenant should alter his farming system as he proposes. I find the tenant is a competent tenant, above average, and who is practising a system of farming suitable for this holding having regard to its character, soil, situation and the fixed equipment available. However, it should be possible for the enterprise to be run by the tenant and one workman.

3.4 **Related earning capacity**

Evidence

Budgetary evidence was given on the related earning capacity by Mr Knoall for the landlord based on a system of growing and keeping:

165	acres winter wheat
60	acres winter barley
40	acres spring barley
20	acres spring beans
80	beef stores (on 40 acres)

This budget left a margin, after deducting costs of £69,210.

This proposed system and gross margin was disputed by the tenant as unrealistic and unsuited to the holding, taking into account his knowledge of the farm. The total fixed costs estimated by the landlord were also disputed as being far too low. The margin on the landlord's proposed system was estimated by the tenant to be £64,305.

Budgetary evidence on the related earning capacity was given for the tenant by Mr Littleman, based on the current system of growing and keeping:

145	acres winter wheat
100	acres winter barley
40	acres spring barley
35	acres potatoes
80	acres pasture

He considered the above system and keeping 400 ewes, finishing 100 barley bulls and keeping 500 root tegs to be the most appropriate and profitable leaving a margin, after deducting costs, of £58,620.

The landlord challenged this system and also cost estimates, in particular labour costs.

Findings

Having inspected the holding and considered the evidence given, I find the tenant's farming system the most appropriate and profitable for this holding. I reject the landlord's proposals as not realistic or acceptable, having regard to the fact that farming is a long-term business. I further accept that in the interests of good husbandry being as much of the land is light soil, sheep have to be kept on the holding. I prefer the tenant's budget to the landlord's.

3.5 **Tenant's improvements**

Evidence and submissions

Evidence was given by the tenant that he had carried out the following improvements at his own expense at the property.

(a) erected 500 tonne grain store in 2002 — cost £22,000
(b) drained OS No 560, 20.12 acres in 2005 — cost £6,000
(c) installed 3 phase electricity supply in 2002 — cost £4,150
(d) improved the farmhouse by fitting out the kitchen with Aga cooker and provided an Ideal 120,000 BUT boiler and 15 radiators in 2004. Total cost £10,000.

No landlord's consent was obtained for these improvements and he contended they would therefore be classed as tenant's fixtures.

The landlord agreed this work had been done and accepted the costs given, but submitted that all the work carried out in the farmhouse was for the tenant's convenience. He agreed in cross-examination that the remaining work was of benefit to the holding.

Neither party gave any evidence of the apportionment or calculated rental value of these improvements.

Findings

(a) The improvements stated have been carried out by the tenant.
(b) No detailed evidence has been given of the annual estimated rental value of these improvements.
(c) I accept that the house improvements are for the comfort and convenience of the tenant. Nevertheless, I consider they have a value to the holding.
(d) I have disregarded the existence of these improvements and have not taken them into account in my rental Award.

4. **Current value of rents of comparable properties**

Evidence

The landlord's expert valuer and land agent Mr JR Andrews, gave general evidence that he had increased and agreed recently the rents of five other holdings on the estate and these farms ranged from 120 to 530 acres. The average rent agreed was £80.05 per acre, but he stated that Old Hendre was the best farm on the estate and was worth appreciably more than the average. He was also aware of a letting of a 358 acres arable farm in Gloucestershire, 10 miles away in February 2005 let at a reputed £95 per acre. He also cited rentals obtained on two farms let recently on Farm Business Tenancies at an average of £130 per acre. The tenant's expert tenant Mr JT Thomas gave the tone of local rent settlements on the nearby Oldbury Estate where he had recently settled the rents of six farms where the rents were agreed at an average of £70 per acre. Also, he submitted the tenant had taken this subject farm at a very high rent, well above other rents in the area and of those quoted, they were merely catching up.

Findings
(a)   I find that the evidence given was merely a tone of local rental values, but no more than this as I was not asked to inspect any holding and was not supplied with any detailed information on any tenancy.
(b)   While I accept the evidence given of lettings on Farm Business Tenancies as evidence, the lettings are on a different basis to Old Hendre which has to have its rental assessed under the provisions of schedule 2 to the 1986 Act.

5.   **Other relevant factors**
Evidence and Submissions
(a)   The landlord's agent submitted that the rental to be awarded is not confined to the agricultural rent alone and is entitled to take into account the marriage/latent capacity rather than the realised capacity of the holding.
      He cited the case of *Childers (JW) Trustees v Anker* [1996] 1 EGLR (CA 1995) and submitted that since the tenant owned and farmed 75 acres of land adjoining (which formerly was part of the farm and The Aral Estate) marriage/latent value should be added to the rental value of Old Hendre. He also maintained that this value was further increased since the tenant was taking 15 acres of nearby grazing each year from the estate and hopefully this could continue for years. He further developed this point by stating that Old Hendre possessed more than an adequate set of farm buildings — in fact sufficient to serve more than 403 acres and also served the other land owned and farmed by the tenant.
(b)   The tenant disputed this and contended that marriage/latent value was irrelevant to this farm as *Childers (JW) Trustees v Anker* referred to a totally different set of circumstances.

Findings
I find:
(a)   that in the light of the judgment given in the *Childers (JW) Trustees v Anker* case, I am obliged to take into account whether or not there is any relevant marriage/latent value in this case

(b)   that there is in fact marriage value attached to the farm buildings which serve not only Old Hendre, but also the 75 acres immediately adjacent which at one time formed part of the holding and is now owned and farmed by the tenant, as one unit

(c)   there is no marriage/latent value attached to the farmhouse and the land and cottages which form a farming unit of 403 acres, well farmed and standing on their own as a viable farm

(d)   that there is no marriage/latent value in the 15 acre grass keep.

6.   **Decision**

In arriving at the rental properly payable for the holding, I am required by the provisions of schedule 2 to the Act to find the rental payable which shall be the rent at which the holding might reasonably be expected to be let to a prudent and willing tenant. I must, and have taken into account all relevant factors stated in paragraphs 1 to 3 of schedule 2 including the terms of the tenancy (including those related to rent), the character and situation of the holding, including the locality in which it is situate, the productive capacity of the holding and its related earning capacity and the current rents of comparable lettings as determined in accordance with paragraph 1(3) of schedule 2 to the Act. I have disregarded the tenant's improvements and appreciable scarcity of comparable holdings for letting.

I have also considered the judgment given in the *Childers (JW) Trustees* v *Anker* case and concluded that marriage value is attached to the farmbuildings at Old Hendre but not the farmhouse, land and cottages.

Having considered the factors and related these to each other, in the context of the holding as a whole, I have arrived at the decision that the rent properly payable for the holding considered as a whole, from 2 February 2008 is the sum of £32,500 per annum and this rental is payable from 2 February 2008.

Costs

Having, subsequent to the publication of my Interim Award, seen the Calderbank Offers made, without prejudice, save as to costs, namely:

Landlord   £36,750 on 2 January 2008
Tenant   £30,000 on 12 January 2008

I find that my Award is substantially nearer the tenant's offer to settle and accordingly award all the costs of the reference to the tenant as directed in my Final Award.

24.15 Arbitrations other than under the Agricultural Holdings Act

Other disputes arise which are not tenancy matters, eg disputes on a valuation of produce, crops etc, on the sale of a farm. These disputes are not resolved under the agricultural arbitration code, but under the provisions of the Arbitration Act of 1996. Certain arbitrations relating to milk quotas are governed by the Agriculture Act 1986.

Where an issue cannot be resolved the contract will usually provide that the matter shall be referred to an arbitrator and it is usual for the contract to state who appoints the arbitrator.

The Arbitration Act 1996, as in the case of tenancy matters provides disputing parties with an opportunity to agree the procedures to best suit themselves. Section 34 provides both procedural and evidential matters and suggests that it is for the arbitrator to decide such matters, but subject to the rights of the parties to argue any matter. It suggests typical matters to be agreed or laid down:

(a) when and where proceedings take place
(b) submissions
(c) discovery
(d) questions/examinations/cross-examination (it is thought that these will be normal procedure as now practised)
(e) rules of evidence
(f) whether arbitrator obtains legal opinion on matters of fact
(g) whether there should be a hearing or written representations
(h) fixing of timetables.

In section 37, unless otherwise agreed the arbitrator may appoint experts or legal advisors but the parties must be given an opportunity to comment on the advice obtained.

24.16 Independent experts

As an alternative to a reference to arbitration and particularly to save costs, the practice of the parties of appointing an independent expert to determine the issue on hand has grown in recent years. This has much to commend and the CAAV, in particular, encourages the method of dispute resolution.

An independent expert must exercise his own professional expertise and judgment and differs from an arbitrator who acts in a judicial capacity. As there is no obligation governing an independent expert, the independent expert does not have to follow the procedures set out in the Arbitration Act 1996, but is appointed either by agreement of the parties (who agree to accept his decision) or under the terms of an agreement previously entered into, prior to the dispute.

The independent expert must ensure that he complies with the joint wishes of the parties, acts promptly and should observe the principles of natural justice in that he must be, and be seen to be disinterested and unbiased and also see that each party should be given a fair opportunity to present his case and know and meet the opposing case.

It is very important at the start that the independent expert settles with the parties his contract, in particular procedure, whether any legal assistance is to be sought, what documents and witnesses are to be produced (he has no power to compel discovery or attendance of witnesses), and also agreement who pays his fees.

There is no right of appeal against his determination. However, he can be liable for any losses sustained by a party arising from his negligence.

24.17 Procedure

24.17.1 Appointment

The appointment of an independent expert may be made variously, by a person nominated in the original tenancy agreement, by written agreement of the parties, the President of the CAAV/RICS or possibly the chairman of the local branch of the CAAV, RICS or Law Society.

The name and address of the parties, exact nature of the dispute and the question to be resolved should be contained in the written appointment. Also, how costs are to be met.

An expert should consider carefully, prior to accepting an appointment, if the dispute falls within his/her expertise.

It may be that procedures to be followed are contained in the original agreement.

Once an appointment is accepted, the independent expert will suggest the further procedure including:

(a)   if a hearing is to be held or not — usually no hearing is held
(b)   time scale for submitting statements of case or written submissions
(c)   time scale for replies or counter-submissions to be made by each party to the opposing case

(d) provisions (if considered necessary) for the independent expert to obtain any specialist advice, eg legal advice and who is to pay for this

(e) possibly an exclusion of liability clause in accepting the appointment — this is strongly recommended and needs very careful drafting

(f) date for inspection and whether the expert will inspect on his own or requires to be accompanied. If to be accompanied, representatives of both parties should be present.

24.17.2 Determination

The decision of the expert is called a determination. This is best in writing but there is no set form. There is no time-limit on the expert to produce his determination, but it is always wise not to unduly delay the issue of the determination.

The parties cannot request a statement of reasons from the independent expert.

Costs are asked to be paid to the expert prior to the issue of his determination. The expert does not have power to award costs unless specifically empowered and even then they should be awarded judicially which means they should follow the event.

Example notification letter

To: Messrs Royston & Harris   Messrs Watts & Smith
    15 The Avenue   1A The Grange
    Hereford   Ross-on-Wye

Dear Sirs

MARSTON HALL FARM, ROSS-ON-WYE

My determination in this matter has now been made and may be taken up when I have received the total amount of my costs which are £2,021.76 (inclusive of VAT) and which you have agreed shall be divided equally between you at £1,015.88 each.

Yours faithfully
RF Russell

Example determination

Name of Holding: Marston Hall Farm, Ross-on-Wye, Herefordshire
Agreement: Tenancy agreement dated 1 November 1995
Landlord: Mr John Astley Harris
Tenant: Mr Archibald Thomas

I Richard Frederick Russell, FRICS, FAAV was on 23 October 2007 appointed in writing by the parties as the Independent Expert to determine in accordance with section 13 of the Agricultural Tenancies Act 1995, the rent payable with effect from 1 November 2008 for the above holding held by the Tenant from the Landlord under the above tenancy agreement. I hereby determine the rental payable at £21,000 (twenty one thousand pounds) per annum.

Dated this 30th day of October 2007
RF Russell FRICS FAAV
of Russell and Jones
15 Church Walk, Ross-on-Wye

Grateful acknowledgement is made to the CAAV for the use of their Guidance Notes for the conduct of the Independent Expert Role, and in particular their Secretary and Advisor, Mr J Moody, Coleford, Glos, who has kindly given much advice on arbitration and other legal issues.

Milk Quotas

25.1 This chapter is concerned only with that milk quota legislation which relates specifically to milk quota registered in the name of a tenant of an agricultural holding.

Many years have lapsed since the introduction of the milk quota system into the United Kingdom on 2 April 1984 and while this section focuses on the legal mechanics it also has regard to how milk quota issues are generally dealt with in practice.

25.2 Compensation

After the introduction of the milk quota system in 1984 it soon became apparent that quota had acquired a value and that a tenant of an agricultural holding with quota should, on termination of the tenancy, be able to share in the value of the quota. Section 13 of the Agriculture Act 1986 gave the right to tenants to receive compensation for milk quota in accordance with the provisions set out in schedule 1 of the Act.

The value of milk quota rose from less than 20 pence per litre in the early years to peak at in excess of 60 pence per litre in the years after the deregulation of the milk industry in 1994. More recently values have eroded down to less than 1 pence per litre as the economics of milking have resulted in the contraction of the UK dairy herd and production falling some distance short of the national quota.

The reduced value of milk quota has resulted in the provisions for the division of the value becoming less important and a tendency for the parties to do deals without recourse to going through the detailed calculations as set out in this section. Such deals avoid the cost of such calculations which erode the value still further. Given that much of the detail for an accurate calculation relates to the base year of 1983, now beyond the memories of most, the collation of information can be a lengthy and onerous task.

While the milk quota is generally in the name of the tenant it must be noted that it is not owned by the tenant. By virtue of the Agriculture Act 1986 the tenant is entitled to receive compensation for his share of the milk quota upon the termination of the tenancy. In practice, because of the decline in the fortunes of the dairy sector many tenants have terminated production (with landlord's consent where necessary) without terminating the tenancy and the quota has been sold off the holding with the interested parties cooperating with the sale and agreeing the apportionment of the proceeds. This process has been further hastened by the amendment of the Dairy Produce Quotas Regulations precluding non producers from holding quota.

It should be noted that holding in the Agriculture Act 1986 has the same meaning as in the Dairy Produce Quotas Regulations, that is 'all the production units operated by the producer and located within the geographical territory of the Community'. In this chapter the expression 'agricultural holding' means that area of the producer's holding which is subject to a tenancy in respect of which:

(a) a claim is to be assessed under the Agriculture Act 1986 section 13 (compensation to outgoing tenants for milk quota)

(b) there is a reference under section 12 of the Agricultural Holdings Act and the rental value of transferred quota falls to be considered under the Agriculture Act 1986 section 15 (rent arbitrations: milk quota).

25.3 The aim of the provisions in schedule 1 is to entitle tenants whose tenancies are terminating to receive compensation for the value of the following quota.

(a) Transferred quota (TQ)

Quota that has been acquired from another producer and transferred to the agricultural holding to the extent that the tenant has borne the cost of the transaction which effected the transfer.

(b) Tenant's excess quota (TE)

All quota which, after deduction of TQ, exceeds the standard quota (SQ) for the agricultural holding. This reflects the extent to which the tenant's management resulted in more quota being allocated to the agricultural holding than SQ.

(c) Tenant's fraction (TF)

A proportion of SQ to reflect the contribution made by the tenant towards the provision of dairy improvements and fixed equipment.

25.4 The steps in outline

In order to prepare a claim it will assist a valuer to break the claim down into a number of steps as follows.

(1) Does the tenancy qualify?
(2) Does the tenant qualify?
(3) Calculate the number of litres of quota registered in the tenant's name which are relevant to the agricultural holding by apportionment if necessary to give relevant quota (RQ).
(4) Calculate how much of the relevant quota is relevant TQ.
(5) The balance (RQ – TQ) will be relevant allocated quota (AQ).
(6) Assess the relevant number of hectares (RH).
(7) Consider whether the use of the reasonable amount (RA) is appropriate.
(8) If the answer is:
  (a) no, select the appropriate prescribed quota per hectare (PQH) and multiply PQH × RH to give SQ or
  (b) yes, assess the RA and select the appropriate PQH and prescribed average yield per hectare (PAYH) from the current Milk Quota (Calculation of Standard Quota) Order and calculate standard quota (SQ).
(9) Calculate the TE quota.
(10) Assess the annual rental value (ARV) of qualifying tenant's dairy improvements and fixtures.
(11) Assess the proportion of rent paid (PRP) during the relevant period in respect of areas used for feeding, accommodating and milking dairy cows.
(12) Calculate TF.

(13) Assess the proportion of TQ (from 4) which qualifies for compensation.

(14) Calculate the number of litres for compensation (QFC).

(15) Assess the value of quota and total compensation due.

25.5 The steps in detail

The steps in detail are as follows.

(1)   Does the tenancy qualify?

Qualifying tenancies are:

(a) a tenancy under which the tenant enjoys security of tenure by virtue of section 2 of the Agricultural Holdings Act 1986

(b) a tenancy for a term of years approved by the minister.

See Agriculture Act 1986 schedule 1 paragraph 18(1) definition of tenancy.

Excluded tenancies are:

(a) a tenancy for a period between one and two years, a *Gladstone* v *Bower* tenancy — now of historic interest only

(b) a grazing or mowing tenancy for a period of less than one year — again of historic interest only

(c) farm business tenancies under the Agricultural Tenancies Act 1995.

(2) Does the tenant qualify?

(a) The tenant must have quota registered as his. Many tenants farm their holdings through the medium of a partnership or a company and the registered producer is not the tenant himself. Such arrangements may disqualify a tenant from claiming compensation and valuers should advise that a tenant's name appears on the register before a claim is made. In practice seldom is this detail made to be an issue.

(b) The quota must be registered in relation to land of which the agricultural holding either is the whole or a part. While in practice this is unlikely to cause difficulties, tenants whose registered address is remote from the agricultural holding should be aware of the requirement.

(c) The tenant must have had quota allocated to him, in relation to the agricultural holding.

(d) If the tenant is claiming compensation for quota transferred at his expense and registered in his name, he may claim compensation in respect of that quota provided that he either:
   (i) had quota allocated to him as in (c) or
   (ii) was in occupation of the agricultural holding as a tenant (though not necessarily under the present tenancy) on 2 April 1984.

See Agriculture Act 1986 schedule 1 paragraph 1(1).

Other tenants who are entitled to claim compensation are:

(e) Tenants who have succeeded to tenancies by virtue of the grant of a tenancy after 2 April 1984:
   (i) following a direction by the Agricultural Land Tribunal under section 39 of the Agricultural Holdings Act 1986 (succession on death) or section 53 (succession on retirement) or under corresponding previous legislation
   (ii) following a direction under section 39 of the Agricultural Holdings Act 1986 where a new tenancy is granted by agreement between landlord, tenant and potential statutory successor to that successor
   (iii) to a tenant who was a close relative of the previous tenant by agreement with the landlord under section 31(l)(b) or (2) of the Agricultural Holdings Act 1986.

See Agriculture Act 1986 schedule 1 paragraph 2.

(f) Tenants who are assignees, where a qualifying tenancy has been assigned after 2 April 1984 (whether by deed or by operation of law). Any allocated quota or transferred quota which would fall to be assessed for compensation in the hands of the assignor should be treated as having been allocated or transferred to the assignee.

See Agriculture Act 1986 schedule 1 paragraph 3.

(g) In the case of sub-tenancies, a head tenant, who sublets to a sub-tenant who is entitled to claim compensation from the head tenant, can claim a like amount from the head landlord.

See Agriculture Act 1986 schedule 1 paragraph 4.

Tenants who may not claim compensation will include those who took occupation of the agricultural holding after 2 April 1984 and did not have quota allocated to them. Thus tenants who have transferred quota or have paid a premium to a landlord for quota and have taken occupation after 2 April 1984 are barred from statutory compensation.

(3) Establish the RQ

(a) Where the agricultural holding is the only land farmed by the tenant, then registered quota will be the RQ.
(b) Where the agricultural holding forms only part of the land farmed by the tenant, the registered quota must be apportioned to the termination date between agricultural holding and the balance of the land farmed.

The procedures for apportionment are set out in the Dairy Produce Quotas Regulations. Briefly, milk quota should be apportioned to areas used for milk production in the last five-year period during which milk production took place before the date of quitting.

Since all those persons who have an interest in the land farmed by the tenant have a right to make representations to an arbitrator appointed to determine an apportionment, both landlord and outgoing tenant should ensure that all interested parties agree that the apportionment is undertaken taking into account the areas used for milk production as specified in the transfer form. Failure to do so may trigger off arbitration.

In practice, a tenant will be well advised to follow the procedures leading to prospective apportionment not more than six months preceding the termination of the tenancy. This should avoid an arbitrator appointed to determine compensation being delayed by having to determine apportionment which may have to be determined at another arbitration by a different arbitrator appointed in accordance with the Dairy Produce Quota Regulations.

The expression 'areas used for milk production' was brought to the attention of the High Court in *Puncknowle Farms Ltd* v *Kane* [1985]

2 EGLR 8. The court ruled that the expression included the following:

> the areas of the set of buildings and yards belonging thereto used for the production of milk and forage areas used by the dairy herd and to support the dairy herd by the growing of grass and any fodder crop for the milking dairy herd, dry cows and all dairy following female young stock (and home bred dairy or dual purpose bulls for use on the premises, if applicable) if bred to enter the production herd and not for sale. In this case, maize, silage, hay and grass were the fodder crops, but consideration would have been given to corn crops, or part of corn crops, grown for consumption by the dairy herd or young stock, including the use of straw, had agreed evidence been produced.

This very wide definition applies only for the purposes of apportionment under the Dairy Produce Quota Regulations and must not be confused with other similar expressions used in the Agricultural Act 1986.

See Agriculture Act 1986 schedule 1 paragraphs 1(1) and 1(2) for definition of the relevant quota. Dairy Produce Quotas Regulations 1997, as amended. Regulation 8 (Apportionment of quota), Regulation 9 (Prospective apportionment of quota) and schedule 2 (Apportionment and prospective apportionments by arbitration).

(4) Calculate how much of RQ is relevant TQ

TQ and AQ are apportioned to the agricultural holding on the same basis. A tenant cannot, therefore, seek to retain all TQ on that part of the land he farms which is not the agricultural holding in respect of which he is claiming compensation.

Since the original allocation of milk quota the national quota and the allocations awarded to produces have been affected by both cuts and additions. As a starting point it is necessary to determine the net impact of such changes on the quota transferred in by the tenant. The calculations for such adjustments are not straightforward however, practitioners have developed calculators incorporating all the changes. By way of example a transfer in of 100,000 litres in the 1984–1985 milk year gives rise to a net figure as at the 2007–2008 milk year of 90,481 litres and further increments are planned and expected beyond 2008–2009 until the probable termination of the scheme in 2014–2015.

(5) *Assess relevant AQ*

Deduction of the relevant TQ from RQ will give the relevant AQ.

(6) *Assess the relevant number of hectares (RH)*

The definition of 'the relevant number of hectares' in the Act is:

    (a)   the average number of hectares of the land in question used during the relevant period for the feeding of dairy cows kept on the land, or

    (b)   if different, the average number of hectares of the land which could reasonably be expected to have been so used (having regard to the number of grazing animals other than dairy cows kept on the land during that period).

The relevant period is the period in relation to which allocated quota was determined. In most cases wholesale quota was allocated on the basis of the number of litres delivered to the Milk Marketing Board (or other purchaser) during the period 1 January 1983–31 December 1983.

Since it is mandatory to use the second definition (b) if it is different to the first definition (a), it is logical to ignore (a) and proceed with (b). If (b) is the same as (a) so be it!

The expression 'land in question' refers to the agricultural holding in respect of which compensation is being claimed.

The reference to 'land used for the feeding of dairy cows kept on the land' is further defined as excluding 'land used for growing cereal crops for feeding to dairy cows in the form of loose grain'. This curious definition is unclear but is possibly meant to exclude land used for growing cereals which were used under definition (a) or could reasonably be expected to have been used under definition (b) for home mixed concentrates. The land used for feeding dairy cows will clearly include grazing land, land used for growing fodder crops for conservation and areas where cows were fed, such as silage clamps and parlour standings with feeders. Other areas which it may be argued fall to be considered are roadways, tracks and other areas which facilitate feeding.

Dairy cows are defined as 'cows kept for milk production (other than uncalved heifers)'. A qualifying dairy cow starts as a newly calved heifer and finishes once she has completed her final lactation. Thus a barren cow kept by a tenant after she has ceased milk production to put on weight for slaughter cannot qualify as a dairy cow but will fall to be considered under definition (b) as a 'grazing animal other than a dairy cow'.

The reference to 'kept on the land' is unclear. There are instances where no dairy cows were in fact ever physically present on the agricultural holding during the relevant period but fodder was taken from the agricultural holding to feed dairy cows kept on other land farmed by the tenant. There are many cases where the agricultural holding is bare land serviced by fixed equipment elsewhere. In such circumstances valuers should bear in mind definition (b) and consider the average number of hectares which could reasonably be expected to have been used during the relevant period for the feeding of dairy cows kept on the land. An objective assessment of the fixed equipment and the land comprised in the agricultural holding may suggest that dairy cows might reasonably have been expected to have been kept on the land where there was the fixed equipment to support them. On the other hand where there was no fixed equipment, then it may be argued that it would not have been reasonable to expect a tenant to have used the agricultural holding for feeding dairy cows.

Definition (b) requires a valuer to have 'regard to the number of other grazing animals ... kept' on the agricultural holding. The interpretation of this subjective test can be argued in at least two ways.

1. Assess the average number of hectares (a) which could reasonably have been expected to have been used for feeding dairy cows. Discover the actual number of other grazing animals kept on the agricultural holding during the relevant period and assess the average number of hectares (b) that might reasonably have been expected to have been used during the relevant period for feeding that number. Area (a) – area (b) will be the relevant number of hectares or

2. Discover the actual stocking rates during the relevant period and use these as a guide to assessing the potential stocking rates when assessing the area which could reasonably have been expected to have been used for feeding dairy cows during the relevant period.

The valuer is required to assess the average number of hectares during the relevant period. This requires a calculation of the areas of the holding used for other purposes than feeding dairy cows and the periods for which they were so used.

Areas which are subject to mixed stocking can be assessed on a livestock unit basis. Areas which are shut up should be considered on the basis of their next use. Thus an area of grass shut up for cutting for silage or feeding cows would be included while shut up. It may be argued on the one hand that the use of keeper sheep in the winter months should be included on the basis that the purpose of the sheep is to improve grassland for the spring growth for feeding dairy cows, and on the other hand it may be argued that the use of the area is for feeding sheep and not dairy cows.

It is convenient to summarise the definition of the relevant number of hectares as:

> the average number of hectares of the agricultural holding which could reasonably be expected to have been used during the period, in relation to which allocated quota was determined, for the feeding of dairy cows kept on the holding (an objective test), having regard to the number of other grazing animals kept on the holding during that period (a subjective test).

See Agriculture Act 1986 schedule 1 paragraph 6(1)(a) 'the relevant number of hectares' paragraph 8 'the relevant period'.

(7) Consider whether the use of the RA is appropriate

The valuer must decide whether to use the calculation set out in the Agriculture Act 1986, schedule 1 paragraph 6 at sub-paragraph 6(1)(b) or at sub-paragraph 6(2).

The circumstances whereby an arbitrator decided to follow sub-paragraph 6(1)(b) were the subject of the decision in *Surrey County Council v Main* [1992] 1 EGLR 26. The judge gave guidance as to how an arbitrator should make his decision whether to follow sub-paragraph 6(2) and use the reasonable amount. The judge said that if an arbitrator decides that the quality of the land and the climatic conditions were not so unorthodox that they would have affected milk yield either way then he need consider sub-paragraph 6(2) no further. If, however, the arbitrator decides that because of either the quality of the land or the climatic conditions in the area, or both, during the relevant period, the amount of milk which could reasonably be expected to have been produced from one hectare of the agricultural holding during the relevant period (RA) differs from the PAYH, the judge said that the calculation set out in paragraph 6(2) must be used.

It is submitted that the quality of the land includes consideration of a wide range of matters from the fixed equipment to the soil. The climatic conditions in the area during the relevant period may relate to unorthodox weather during the relevant period, such as a drought or a flood, or may relate to the agricultural holding being located in an area with an unorthodox weather problem, such as a rain-shadow.

(8) Calculate the SQ

The SQ is intended to reflect what a reasonable landlord would expect a reasonable tenant to have in the way of milk quota allocated to the holding having regard to the type of farm and the fixed equipment provided. If the AQ is greater than the SQ this is due to the tenant's good management from which the tenant derives recognition with the excess while, conversely if the allocated quota is less than the SQ the tenant's fraction has to be reduced proportionately to reflect the shortfall.

For the calculation of the SQ either:

(a) If the use of the RA is not appropriate select the correct prescribed quota per hectare (PQH) and calculate SQ

Multiply the RH by the appropriate PQH.

PQH is that number of litres which the minister (to take account of national adjustments to registered quota) prescribes as being appropriate for three different classes of dairy cows and three different land types.

$$SQ = RH \times PQH$$

or

(b) If the use of the RA is appropriate, assess RA, select the correct PQH and PAYH and calculated SQ.

(i) Prescribed PAYH is that number of litres which the minister has prescribed as being appropriate for the same three different classes of breeds of dairy cow and three different land types as for prescribed quota per hectare. The average yields are those applying to the relevant period and are set out in the current Milk Quota (Calculation of Standard Quota) Order.

(ii) Valuers should note the arbitrary dates on which amendments to the Milk Quota (Calculation of Standard Quota) Order come into effect. They do not coincide with either the quota year or most traditional term dates.

(iii) The formula for assessing standard quota is:

$$SQ = RH/PAYH \times RE \times PQH$$

where:

RH   is the relevant number of hectares

PAYH is the prescribed average yield per hectare

RA   is the reasonable amount

PQH   is the prescribed quota per hectare

See Agriculture Act 1986 schedule 1 paragraph 6(2) 'the reasonable amount'.

(9) Calculate TE quota

The TE is derived by deducting the SQ from the AQ. The result is the TE which qualifies for compensation.

In the event the AQ is less than the SQ the tenant's fraction calculated at a later point has to be reduced by the percentage of the shortfall.

See Agriculture Act 1986 schedule 1 paragraph 6(3).

(10) Assess the PRP

The ARV is the annual rental value at the end of the relevant period of the tenant's dairy improvements and fixed equipment on land used for the feeding, accommodating or milking of dairy cows kept on the agricultural holding. The annual rental value is that amount which would fall to be disregarded under paragraph 2(1) of schedule 2 to the Agricultural Holdings Act 1986 on a reference to arbitration as to rent properly payable as at the end of the relevant period.

A valuer will therefore sieve each item which he has to consider as follows.

(a) Did it exist at the end of the relevant period?

(b) Was it on land actually used at some time during the relevant period for feeding, accommodating or milking of dairy cows?

(c) Was it a dairy item?

(d) Was it fixed equipment as defined in section 97 to the Agricultural Holdings Act 1986? It is immaterial for these purposes whether it is also a tenant's improvement.

(e) Did it contribute to an increase in rental value of the holding?

(f) Was there an obligation to provide it imposed on the tenant by the terms of the contract of his tenancy?

(g) What would the rent properly payable for the agricultural holding in accordance with paragraph 1 of schedule 2 to the Agricultural Holdings Act 1986 (that is including all tenant's improvements and fixed equipment) have been at the end of the relevant year?

(h) By how much should the rent assessed under (g) be reduced by reason of the disregard of increase in rental value of the item? That amount will be the ARV.

See Agriculture Act 1986 schedule 1 paragraph 7 'tenant's fraction'. Agricultural Holdings Act 1986 schedule 2 paragraph 2(1) and section 97 definition of fixed equipment.

(11) Assess the proportion of rent paid (PRP)

Whereas ARV is assessed on an arbitrated basis at the end of the relevant period, PRP is carved out of the rent actually paid for the agricultural holding in respect of the relevant period attributable to those parts of the agricultural holding used for the feeding, accommodating or milking of dairy cows. Thus a benevolent landlord who was letting at a low rent will be at a disadvantage to a tenant who had just completed a dairy improvement scheme two days before the end of the relevant period: see the decision in *Creear* v *Fearon* [1994] 2 EGLR 12.

(12) Calculate the TF of SQ

TF is ARV/ARV + PRP

(13) Assess the proportion of TQ which qualifies for compensation

If the tenant has borne all the cost of the transaction, 100% of the transferred quota will qualify for compensation; if, say, only 50% then only 50% of the transferred quota will qualify for compensation.

See Agriculture Act 1986 schedule 1 paragraphs 1 and 5(3).

(14) Calculate the total number of litres of quota for which the tenant is entitled to compensation (QFC)

(a)   Where AQ is greater than SQ:
$$QFC = TQ + [AQ - SQ] + [TF \times SQ]$$

(b)   Where AQ is less than SQ:
$$QFC = TO + [TF \times AO \times AO/SQ]$$

(15) Assess the value of quota and total compensation due

The value of milk quota which qualifies for compensation is

> the value of the milk quota at the time of the termination of the tenancy in question and in determining that value at that time there shall be taken into account such [all] evidence as is available, including [but not necessarily exclusively] evidence as to the sums being paid [at that time] for interests in land in cases where
>
> (a)   milk quota is registered in relation to the land; and
> (b)   no milk quota is so registered.

The issue of the value was a major one particularly when open market trade was being undertaken at prices over 30/40 pence per litre and such trades were being conducted on the back of short term tenancy arrangements. Since most agricultural tenancies preclude the granting of sub tenancies there was an argument that a tenant could not in practice realise the value of quota apportioned to the holding.

More recently the regulations have been amended so that without land transfers now make up the vast majority of quota sales which avoids the need for any tenancy arrangements. Determining the value quota is now a relatively simple task.

Practitioners will find Dairy Co (formely the Milk Development Council) a useful source of information. Dairy Co collates details of quota sold by the principle quota agents in the UK and make the data available.

See Agriculture Act 1986 schedule 1 paragraph 9.

Procedural and other matters
25.6 Establishing SQ and TF during a tenancy

Where a tenant may be entitled to compensation for quota at the termination of his tenancy, either the landlord or tenant may at any time before the termination of the tenancy by notice in writing served on the other demand that the determination of the standard quota for the land or the tenant's fraction shall be referred to arbitration.

Both tenants and landlords may wish to know standard quota and tenant's fraction during a tenancy. From the tenant's point of view it will enable him to have a clear indication of the number of litres which will qualify for compensation at termination. The procedure has the distinct advantage of ensuring that at termination he will receive compensation sooner than if standard quota and tenant's fraction have to be referred to arbitration after termination. From the landlord's point of view he will have a clear idea of how many litres he is likely to have to pay for and in the event of his re-letting the farm, he can decide how he will deal with the incoming tenant. In that both standard quota and tenant's fraction require consideration of historical facts during the relevant period, the procedures will assist both tenant and landlord by dealing with the questions before facts are lost in the mists of time. In any event, both tenant and landlord will be assisted by agreeing historic factual information even if that agreement is not crystallised by an agreement as to standard quota and tenant's fraction.

There is provision whereby any changes in circumstances between the time of any agreement in writing as to standard quota or tenant's fraction (or a determination by arbitration) and the termination of the tenancy can be taken into account by an arbitrator appointed to determine compensation at termination, provided that the circumstances are materially different.

Such circumstances will include a reduction in the area of the agricultural holding.

See Agriculture Act 1986 schedule 1 paragraph 11 (6) and (7).

25.7 Notice of intention to claim

Written notice of the tenant's intention to make a claim under section 13 Agriculture Act 1986 for compensation for milk quota must be served on the landlord within two months of the termination of the tenancy. Failure to serve such a notice is fatal and any subsequent claim is unenforceable.

See Agriculture Act 1986 schedule 1 paragraph 11 (1).

There are, however, special rules applying where an application has been made for a direction following the death of a tenant and to circumstances where a tenant remains lawfully in occupation of the holding after termination of his tenancy. Such special rules are outside the scope of this chapter.

See Agriculture Act 1986 schedule 1 paragraphs 11 (3) and (4).

25.8 Time-limit for settling by agreement

Landlord and tenant have eight months from the termination of the tenancy to settle the claim in writing. After eight months, the claim must be referred to arbitration. It is submitted that even if landlord and tenant agree a settlement after eight months they will nevertheless require a consent award by an arbitrator.

See Agriculture Act 1986 schedule 1 paragraph 11(2).

25.9 Arbitration procedures

Arbitration must follow the code set out in schedule 11 of the Agricultural Holding Act 1986 except that the maximum period by which compensation and costs are to be paid after the delivery of the award is extended from one month to three months. There is no provision for the payment of interest.

See Agriculture Act 1986 schedule 1 paragraph 11(5).

Valuers acting for tenants in protracted disputes at arbitration should always bear in mind that an arbitrator may make an interim award, if he thinks fit.

See Agricultural Holdings Act 1986 schedule 11 paragraph 15.

An example

Base facts

Richard Nightingale is tenant of two agricultural holdings and has been for since before 1984. The holdings comprise:

(a)   Hillview Farm extending to 60.5 hectares owned by William Buzzard. The farm has a modern dairy unit with accommodation for 120 Friesian cows and 30 young stock units.

(b)   Blackacre extending to 10 hectares permanent pasture owned by the County Council. The tenant has used the land exclusively for the grazing of heifers.

The quota registered in the name of the tenant comprises 746,248 wholesale litres within which there is 100,000 litres that was purchased at the sole expense of the tenant in the 1987–1988 milk year.

The relevant period was the calendar year of 1983 — the normal base year.

Mr Nightingale has decided to retire and gave both his landlords notice to quit to take effect at Michaelmas 2007.

The process for calculating the tenant's entitlement to compensation is as follows.

(1)   **Tenancies**
      Both are qualifying agricultural holdings dating back to pre-1984.

(2)   **The tenant**
      Mr Nightingale is a sole trader who has the milk quota registered in his own name.

(3)   **The RQ for each holding by apportionment**
      Hillview Farm: The areas used for milk production over the five years prior to the surrender date (Michaelmas 2007) have averaged 60 hectares per annum.
      Blackacre: The entire 10 hectares has been grazed for the five years.

      Apportionment of quota is therefore:

Hillview Farm	60/70 × 746,248 litres =	639,641
Blackacre	10/70 × 746,248 litres =	106,607
		746,248 litres

(4)   **The relevant TQ**

Transferred quota (1987–88) milk year	100,000	litres
By calculation this equates to	96,420	litres

	Therefore TQ per holding		
Hillview Farm	60/70 × 96,420 litres =		82,646
Blackacre	10/70 × 96,420 litres =		13,774
			96,420 litres

(5) **The AQ**

Therefore AQ per holding

	RQ	TQ	AQ
Hillview Farm	639,642 –	82,646 =	556,995
Blackacre	106,607 –	13,774 =	92,833

Summary			
Total AQ		=	649,828 litres
Total TQ		=	96,420 litres
Total registered quota		=	946,248 litres

((6) **The RH**

The base year is 1983. During 1983 Mr Nightingale used some 50 hectares for feeding dairy cows and 10 hectares for feeding an average of 30 heifer yearlings and 30 heifer calves.

An assessment of Hillview Farm suggests that the fixed equipment of the farm is capable of supporting a dairy herd of 120 cows and that it would be reasonable to have expected that an average number of hectares required to feed 120 dairy cows during 1983 would have been 48.5 hectares, ie RH = 48.5. This would leave some 11.5 hectares available for the young stock which were kept on the holding during 1983.

Blackacre:
During the relevant period no dairy cows grazed at Blackacre and no silage or hay to feed dairy cows was taken from the land. As an overview given location and land type it would not have been reasonably expected that the land would have been used by the dairy cows. The RH is therefore nil.

Summary of RH	
Hillview Farm	48.5 hectares
Blackacre	0.0 hectares

(7) **Consider whether the use of the RA is appropriate**

There is no evidence to suggest that either the quality of Hillview Farm or the climatic conditions in the area during 1983 were so unorthodox as to have affected milk yield, and the use of RA is not appropriate.

(8) **Calculate SQ**

The appropriate prescribed quota per hectare (PQH) figure for Friesian ('other') cows kept on 'other land' is found in the Milk Quota (Calculation of Standard Quota) (Amendment) Order 1992 (SI 1992 No 1225).

	PQH	×	RH		SQ	
Hillview Farm	7,140		48.5	=	346,920	litres
Blackacre	7,140		0.0	=	0	litres

(9) **Calculate TE quota**

	AQ	−	SQ		TE	
Hillview Farm	556,995		346,290	=	210,705	litres
Blackacre	92,883		0	=	92,883	litres

(10) **ARV of the tenant's dairy fixtures and improvements**

Because Blackacre has no RH there is no SQ and therefore no need to make any assessment for the tenant's fraction since this will also be nil. The remainder of the calculations therefore relate to Hillview Farm only.

As at 31 December 1983 the tenant's dairy improvements and fixtures on land used for feeding and accommodating dairy cows were:

(i)   cubicles for 120 cows
(ii) slurry tower
(iii) all parlour fittings, feeders and milking equipment
(iv) bulk feed bin built in the roof of the parlour.

The rent properly payable for the holding for the base year of 1983 excluding the items listed above was £6,150 but had the items not been his he would have been prepared to pay £7,150. The ARV is therefore the difference (ie £1,000).

(11) **Assess the proportion of rent paid in respect of land used for feeding, accommodating or milking dairy cows (PRP)**

During 1983 Mr Nightingale paid a total rent of £5,250.00 for Hillview Farm. The land not used for feeding, accommodating or milking dairy cows included the farm house and garden, the farm cottage, and land and buildings used for feeding and accommodating his young stock. It is assessed that the rent should be apportioned £4,000.00 to the land used by dairy cows and £1,250.00 to the land used for other purposes.

Therefore PRP = £4,000.00

(12) **TF**

$$\frac{ARV}{ARV + PRP} = TF$$

$$\frac{£1,000}{£1,000 + £4,000} = 0.20 \text{ or } 20\%$$

(13) **TQ which qualifies for compensation**
The tenant acquired the TQ at his own expense. All the TQ therefore qualifies for compensation.

(14) **Calculate litres for compensation (QFC)**

Hillview Farm	TE	=	210,705	
	TF	=	69,258	
	TQ	=	82,646	
	QFC	=	362,309	litres

Blackacre	TE	=	92,833	
	TF	=	0	
	TQ	=	13,774	
	QFC	=	106,607	litres

Summary of landlord's apportionment

Derived from	SQ – TF			
Hillview Farm	AQ	=	356,290	
	TF	=	69,258	
Total		=	277,032	litres
Blackacre	SQ	=	0	litres

Overall summary		
Tenant's entitlement		
Hillview Farm		362,609
Blackacre		106,607

Landlord's entitlement		
Hillview Farm		277,032
Blackacre		0
Total registered quota		746,248 litres

(15) **Total compensation due**

Evidence suggests the value of the quota as at 29 September 2007 was 2.5ppl. Compensation due to the tenant is therefore:

Hillview Farm	= 362,609 litres @ 2.5ppl =	£9,065.23
Blackacre	= 106,607 litres @ 2.5ppl =	£2,665.18

25.10 Rent and milk quota

Section 15 of the Agriculture Act 1986 directs an arbitrator, subject to any agreement between the landlord and tenant, to disregard any increase in rental value of the agricultural holding due to relevant transferred quota which would fall to be apportioned to the agricultural holding on a change of occupation.

At an arbitration under section 12 of the Agricultural Holdings Act 1986, once an apportionment of registered quota has been agreed, the only quota subject to the above is transferred quota. Thus quota in excess of standard quota and the tenant's fraction of standard quota which may qualify for compensation on quitting, are not subject to the directions contained in section 15 of the Agriculture Act 1986.

Assuming that the tenant's actual milk quota would be available to a hypothetical prudent, willing and competent tenant, it is submitted that the most satisfactory way to deal with milk quota at rent reviews is to consider the productive capacity (which is a matter of physical output expressed in tonnes or litres) and the related earning capacity in the light of that productive capacity. Related earning capacity will be affected by the number of litres of quota which a tenant has available in relation to the agricultural holding, in that he cannot exploit milk production in excess of that quota without incurring liability to levy.

See Agriculture Act 1986 section 15 and Agricultural Holdings Act 1986 schedule 2 sub-paragraphs 1(1) and (2).

PD Carter MA (Oxon), FRICS, FAAV
Updated by A J Carver MRICS FAAV, 2008

Soil Climate and Productive Capacity

26.1 The quality and productive capacity of agricultural holdings is, to a large extent, governed by its soils and climate. Accurate soil assessment and the ability to interpret soil and climatic data for valuation purposes require training and practice that have often been neglected by chartered surveyors. The aims of this chapter are to identify which soil and climatic data are significant, to describe what published information is available, and to present a system for assessing land quality and productive capacity. Finally, some case histories are presented in which the provision of precise soil and climatic data has played a key role in agricultural valuations.

26.2 Soil maps and agricultural land classification

In the late 1960s and early 1970s the Ministry of Agriculture, Fisheries and Food (MAFF) published a set of Agricultural Land Classification (ALC) maps of England and Wales at a scale of 1:63,360. These maps grade the land according to the severity of environmental constraints on agricultural production, taking into account such factors as soil, gradient, rainfall and altitude. There are five grades, the best being Grade 1 which is land with only very minor limitations, typified by Lincolnshire silt-land. Grade 5, with very severe limitations includes, for example, rough grazing at high altitudes in Wales and the West Country.

ALC maps were primarily intended as a planning tool, to identify and protect the best agricultural land, and this purpose they have served well. However, they are described by Department for the Environment Food and Rural Affairs (DEFRA) as 'provisional' because the amount of fieldwork that went into locating grade boundaries varied considerably from place to place. In some areas the grading was based on detailed ground observations of soil and topography, while elsewhere existing cropping patterns were the main guide. Thus, DEFRA state that parcels of land of less than 80ha cannot be reliably graded using ALC maps alone, and further investigations are necessary. Moreover, the brief descriptive booklets accompanying the maps are insufficiently detailed to give the reasoning behind the grading of a particular piece of land.

Unfortunately, few users of ALC maps (apart from DEFRA officials and planners) are aware of these limitations. Estate and land agents, in particular, have come to use the maps as a definitive statement of land quality in valuations for sale and rental purposes, but in many instances, such use is inappropriate and misleading. The valuer would arrive at a more accurate assessment by obtaining detailed soil, climatic and other site information specific to the land in question. This can then be translated into an accurate assessment of grading if required, using the revised guidelines published by MAAF (Agricultural Land Classification of England and Wales, 1988). Indeed, the ALC grading can be a useful summary of land quality provided it is properly researched and the reasons for a particular grading stated.

The 1988 revised version of the Agricultural Land Classification contains mainly the guidelines and criteria for land classification; the original ALC maps themselves have not been revised. The classification now has five grades plus two subgrades of Grade 3. The most productive and flexible land falls into Grades 1 and 2 and Subgrade 3a and collectively comprises about one-third of the agricultural land in England and Wales. About half the land is of moderate quality in Subgrade 3b or poor quality in Grade 4. The remainder is very poor agricultural land in Grade 5, which mostly occurs in the uplands.

Grade 1 — excellent quality agricultural land

Land with no or very minor limitations to agricultural use. A very wide range of agricultural and horticultural crops can be grown and commonly includes top fruit, soft fruit, salad crops and winter harvested vegetables. Yields are high and less variable than on land of lower quality.

Grade 2 — very good quality agricultural land

Land with minor limitations which affect crop yield, cultivations or harvesting. A wide range of agricultural and horticultural crops can usually be grown but on some land in the grade there may be reduced flexibility due to difficulties with the production of the more demanding crops such as winter harvested vegetables and arable root crops. The level of yield is generally high but may be lower or more variable than Grade 1.

Grade 3 — good to moderate quality agricultural land

Land with moderate limitations which affect the choice of crops, timing and type of cultivation, harvesting or the level of yield. Where more demanding crops are grown, yields are generally lower or more variable than on land in Grades 1 and 2.

Subgrade 3a — good quality agricultural land

Land capable of consistently producing moderate to high yields of a narrow range of arable crops, especially cereals, or moderate yields of a wide range of crops including cereals, grass, oilseed rape, potatoes, sugar beet and the less demanding horticultural crops.

Subgrade 3b — moderate quality agricultural land

Land capable of producing moderate yields of a narrow range of crops, principally cereals and grass or lower yields of a wider range of crops or high yields of grass which can be grazed or harvested over most of the year.

Grade 4 — poor quality agricultural land

Land with severe limitations which significantly restrict the range of crops and/or level of yields. It is mainly suited to grass with occasional arable crops, the yields of which are variable. The grade includes very droughty arable land.

Grade 5 — very poor quality agricultural land

Land with very severe limitations which restrict use to permanent pasture or rough grazing, except for occasional pioneer forage crops.

For planning appeals in relation to development of land for building or mineral extraction it is usually necessary to employ a consultant who understands the ALC system to carry out a soil survey and reassess the grading of the land. DEFRA will normally make its own re-survey where it is seeking to protect high quality agricultural land.

26.3 Soil surveys

The organisation with responsibility for mapping the soils of England and Wales is the Soil Survey and Land Research Centre. The Soil Survey of Scotland operates from the Macaulay Land Use Research Institute, Aberdeen. A soil survey has been established by the Department of Agriculture and Rural Development for Northern Ireland in Ulster.

The whole of mainland Britain is covered by published soil maps at a scale of 1:250,000 (1 inch = 4 miles) with accompanying Regional Bulletins describing the soils and land use. About 10% of England and Wales and most of lowland Scotland and some highland areas are mapped at a scale of 1:63,360 (1 inch = 1 mile) or 1:50,000. A further 15% of England and Wales is covered by detailed maps at 1:25,000 (2.5 inches = 1 mile).

26.4 Collecting soil information

Soil maps are usually made by soil surveyors who are graduates in the field of natural science such as geology or geography. With the aid of an auger and spade the surveyor examines the layers of soil to bedrock or rooting depth (100–120cm). By making a series of such observations across the landscape they are able to produce a map showing the boundaries between different soils. Factors that are significant in making a soil survey are described in detail in the *Soil Survey Field Handbook* and include the following.

26.4.1 Geology

This concerns the kind of parent material in which soils have developed and which usually controls their characteristics. Information about the geology is obtainable from maps produced by the British Geology Survey. Geological maps come in 'Solid' and 'Drift' editions. Solid geological maps show the location of rocks formed in pre-glacial times such as chalk, granite and London clay. In many parts of Britain there is little or no relationship between the soils and the underlying rocks because a succession of glacial and interglacial episodes in the past two million years has blanketed the landscape with drift deposits such as Chalky Boulder Clay, gravel and Brickearth. In many places, the soils have developed in these drift deposits and so it is the drift edition that is the most useful of the two, particularly where no soil maps are available. However, geological maps are not a satisfactory substitute for a soil survey. For example, in some districts the Lower Greensand outcrop on the geological map correctly depicts an area of light soils but these may range from barren sands to some of the best quality light loams.

26.4.2 Colour

The colour of a soil is a reflection of its natural drainage state, and can be used to make an assessment of the risk of waterlogging, even in summer, when the soil is dry. A permeable, well drained soil is a uniform brown or reddish colour throughout. Waterlogging produces grey and orange colours (termed 'gleying') that may range from a few mottles in the case of slight waterlogging to a predominance of grey or blueish-grey colours in severely waterlogged soils. Careful observation is required in soils in reddish parent materials as the grey coloration produced by waterlogging is less pronounced.

26.4.3 Permeability

Soil waterlogging has two principal causes. In low-lying situations with permeable soils such as fenland and river alluvium a groundwater-table rises through the porous subsoil in winter, usually falling again in summer. If adequate outfall or pumping facilities are available this situation can be permanently remedied by arterial drainage, and the agricultural land is then often of the highest quality.

In Britain, however, waterlogging of soils is more commonly caused by the presence of impermeable clay or heavy loam in the subsoil that prevents percolation of rainwater, producing a perched or surface water-table in the upper layer of the soil. Improvement of such soils requires a pipe drainage system with permeable backfill and continuing secondary treatment in the form of moling and subsoiling or some combination of both. This type of waterlogging can never be fully relieved, and care will always be needed in autumn and spring to avoid compaction by stock and machinery.

26.4.4 Texture

This refers to the proportions of sand, silt and clay particles that make up the soil. Texture is usually assessed by moulding moist soil between finger and thumb. Accurate assessment requires training and constant practice. A useful guide to texturing is the DEFRA Leaflet No 895 *Soil Texture*.

26.4.5 Stones

Excessive stoniness reduces the amount of water available to a growing crop, increases implement wear and interferes with drilling and harvesting. In extreme cases, stones and boulders may render land unimprovable.

26.4.6 Depth

Depth to rock, clay, gravel, sand or peat affects root development, drought risk and soil drainage.

26.4.7 Organic matter

Organic matter is a very important soil attribute. Peat, being moisture retentive and easily cultivated, produces some of the best horticultural soils in lowland areas. However, with drainage and cultivation, peat wastes away (at a rate of 1–2cm a year on the Cambridgeshire fens) and so it is a diminishing asset on some farms. In high rainfall areas, a peat top soil is a disadvantage, being often acid and infertile, and because it has a low load-bearing capacity the land poaches easily, resulting in a shorter grazing season compared with adjacent non-peaty soils.

In mineral soils, particularly on light land, continuous arable rotations lead to a reduction in the organic matter content of the topsoil to an undesirably low level, reducing the water-holding capacity and weakening the soil structure so that the surface slakes, caps and forms pans readily. Soil erosion by water may occur under these circumstances.

26.4.8 Calcium carbonate content

The natural calcium carbonate or free lime content of a soil affects its structure and pH level. The content of free lime in a soil is tested in the field by applying hydrochloric acid (10% solution) to a sample of soil. The calcium carbonate content of the Chalky Boulder Clays of East Anglia, such as in the Rodings of Essex, makes them better drained, more manageable and generally higher yielding than soils on acid London, Wealden or Lias Clays.

26.4.9 Soil acidity

Strong acidity (pH less than 4.5) is common in unimproved soils of upland areas. Even lower pHs of between 2.0 and 4.0 may occur in agricultural soils in lowland areas when iron sulphide (pyrites) is present. This sulphide is most common in marine alluvium or peat and so extreme acidity is particularly associated with land in eastern England that is at or near sea level, but investigations have shown that it is not confined to these areas. The problem is caused by the oxidation of sulphide following drainage, and the production of sulphuric acid that causes the sharp fall in pH. At this level of acidity soil material is generally toxic to roots as soluble aluminium becomes available to plants.

Root development is impaired and crops wilt through drought stress on apparently water-retentive soils. Affected soil layers are usually in the subsoil, below the depth of routine lime testing and the problem can be sufficiently severe to reduce the sale or rental value of affected land.

26.4.10 Gradient

The angle of slope affects use of machinery and there are certain critical gradients in the Agricultural Land Classification. For example, a slope of 1 in 5 is the upper limit for satisfactory cultivations and steeper land is placed in Grades 4 and 5.

26.4.11 All the above soil characteristics combine to give recognisable layers or horizons constituting the soil profile. It is the properties and arrangement of these horizons that identify a soil series, which is the common unit of soil mapping in Britain. A soil series consists of soils with similar profile characteristics and formed from similar geological parent materials and so in a similar climate can be expected to have the same management characteristics. A soil series is given the name of the place where it was first recognised or is particularly extensive. For example, Fladbury series identifies a poorly drained alluvial clay first recognised in the Vale of Evesham.

A phase may be used to subdivide a series to emphasise a local characteristic such as topsoil stoniness that is likely to have special management implications.

In Scotland and on small-scale maps in England and Wales, soil associations containing a number of component series, usually in a specified relationship, are the basic mapping unit.

26.5 Collecting climatic data

Soil interacts with climate to affect the productive capacity of a piece of land. Thus, a poorly drained heavy loam of the Brickfield series in Dyfed, receiving 1,400mm of rainfall, will be under permanent or rough pasture with a safe grazing period of only 60 days. A similar soil in Cambridgeshire with 550mm of rainfall will be under continuous cereal rotations.

Agro-climatic data relevant to the assessment of land quality include the following.

(i) Moisture deficit is the water requirement of a crop during the growing season that is not supplied by rainfall in that period. This deficit must be met either from reserves of moisture in the soil, or by irrigation, if yield is not to be affected.

(ii) Accumulated temperature is a measure of heat energy available for plant growth. It is commonly calculated for the period from January to June as the sum of the average daily temperature above 0°C on each day. Values in England range from 850°C on the north Pennines to 1600°C around Bournemouth.

(iii) Field capacity period is a function of rainfall and evapo-transpiration, predicting the length of the period in winter when there is no moisture deficit and the soil is replete with moisture. Waterlogging will occur during this period if permeability and drainage are inadequate, and top soils are liable to compaction by stock and machinery. In parts of Dyfed the field capacity period may last from mid-September to early May, compared with early December to late March in Essex.

(iv) Exposure to strong winds is most detrimental along the coast and at high altitudes where it results in crop damage, particularly to trees. Forestry Commission Leaflet No 85 explains that Topex method of measuring exposure.

26.6 Soil, climate and productive capacity

This section describes how soil and climatic data are brought together to assess productive capacity.

26.6.1 Droughtiness

This is an assessment of the ability of a soil to provide a growing crop with water. Soils vary in the amount of water they can store and release to a crop (known as available water). Medium and coarse sands, for example, can store relatively little water but most of what is stored is easily accessible to a growing crop. Clays, on the other hand, can store a lot of water but it is held at such high tensions in small pores that a large proportion is unavailable to the crop. Fine sands, silts and peats have the largest amounts of available water, which is partly why such soils are renowned for their agricultural quality. An estimate of the water available to crops in a particular soil in the field is made by comparison with a similar soil for which the actual quantity has been measured in a laboratory. Average amounts of available water for common soil series are given in some Soil Survey publications but precise information relating to individual profiles normally requires calculation by a soil specialist.

Because crops vary in their rooting characteristics and efficiency of water uptake, the available water value is adjusted for a particular crop. Thus, in the Hanslope series clay soil the amount of water available to sugar beet is 150mm whereas only 110mm is available to potatoes.

Droughtiness is calculated by subtracting the moisture deficit for a particular crop from the soil available water. A value of more than +50mm indicates that a soil has appreciable reserves of available water in the average season and provides some additional resistance to drought stress in dry years, and is termed non-droughty. A value of between 0 and 50mm indicates a risk of drought stress in dry years and is termed slightly droughty. A negative value of between 0 and 50mm, termed moderately droughty, indicates loss of yield due to drought stress in most years. A value of more than –50mm, termed very droughty, implies severe shortage of water in all but the wettest summers.

This calculation of droughtiness is one of the most significant interpretations of soil and climatic data for it can demonstrate how, because of water shortage, crops in some soils cannot achieve the high yields expected of them, except in a very favourable year. On some land, such as Cotswold brash, this is well known, but elsewhere the significance of drought stress may not be fully appreciated.

The droughtiness calculation also helps with the assessment of the likely benefits of irrigation by demonstrating the drought risk for unirrigated crops. Some marginal soils such as Breckland sands can become highly productive root and vegetable land once water is applied, totally transforming their economic value. However, not all soils are as well suited to irrigation as those of the Breck. Large parts of the eastern counties and Midlands have impermeable clay subsoils in which successful drainage depends upon them drying out sufficiently in summer for deep cracks to form, which will then conduct winter rainfall to mole channels and drains. Good cracks will not form if such soils are irrigated, resulting in wet ground conditions at harvest time.

26.2.2 Trafficability

This is an assessment of the ease with which a soil can be transversed and cultivated and of its susceptibility to structural damage. The critical times for cultivation are in autumn, from September to November, and the spring months of March and April. The days on which land can be worked with little risk of structural damage are called 'potential machinery work days' and these represent an interaction between soil and climate.

The most useful climatic concept in assessing the trafficability of land is the field capacity period — the time during which there is no soil moisture deficit and the soil is replete with water. However, some soils are wet for longer periods than others owing to differences in texture, porosity and topography, and so the climatic data must be modified to represent more closely the likely soil conditions.

The relevant soil factors are those which relate to the ability of the soil to retain or dispose of water, as moisture content affects the bearing strength of the topsoil. These factors are the duration of wetness at various depths, the depth to any horizon which restricts water movement, and the clay content of the topsoil. They are combined to produce an adjustment to the field capacity data. In the case of a well drained, porous loamy soil there are some 20 days in autumn and 10 days in the spring *within* the field capacity period when the land is trafficable. In contrast, for an impermeable clay soil there are some 30 days in autumn and 10 days in spring *outside* the field capacity period when the land is not trafficable.

The effect of soil and climate on machinery work days is demonstrated by soils within the Clifton association in northern England. This soil association is denominated by seasonally waterlogged, impermeable clay loams of the Clifton series, closely associated with lighter and better drained Claverley, Salwich and Quorndon series. The association is mapped near Carlisle with 850mm of rainfall and Darlington with only 600mm.

26.6.3 Crop suitability

The suitability of a piece of land for any crop is a reflection of the soil and climatic factors described in the preceding sections. The number of safe machinery work days in autumn and spring are used to estimate the time available to cultivate the land and to sow and harvest the crop in the right conditions. Droughtiness classes are used to assess the ability of the soils to supply sufficient water for optimum crop growth. For certain crops specific requirements, for example sensitivity to pH, surface stoniness, etc, must also be taken into account.

Four suitability classes — well, moderate, marginal and unsuited — are used and these classes are defined below.

Well suited

Potential production is high and sustainable from year to year. There is adequate opportunity to establish the crop in average years at or near the optimum sowing time and harvesting is rarely restricted by poor conditions. Even in wet years (up to a frequency of about one in four) working conditions are acceptable and do not prevent crop establishment. There are sufficient soil water reserves to meet the average requirements of the crop.

Moderately suited

Potential production is usually moderate or high but can be lower in years when soil water is insufficient to sustain full growth or when crop establishment is unsatisfactory owing to untimely sowing or poor soil structure. In favourable years, which may be wet years on droughty land or dry years in a generally wet climate, outstanding yields can be achieved, but production is not predictable. For root crops, harvesting can be difficult and soil structure may be damaged with consequent penalties for the following crop.

Marginally suited

Potential production varies from year to year. There are considerable risks, high costs, or difficulties in maintaining continuity of output, which are due to the interaction of climate with soil properties, disease or pest problems. In some years there may be a failure to establish the intended crop. Often marginal suitability does not imply high risks in producing the particular crop but rather reflects difficulties in fitting it into a continuous farming system.

Unsuited

Criteria of unsuitability vary with individual crops, but are chiefly climate, gradient and, for root crops, stoniness. Near the selected climatic limits for a crop there will be years with favourable weather which allows efficient production and other years which are too wet or cool. The climatic limits chosen, however, are expected to exclude normal production of the crop at least one year in four: a higher degree of risk was judged unacceptable.

The assessments assume a moderately high level of management and appropriate underdrainage measures. Factors such as market conditions and farming systems are not taken into account.

The following are examples of the method for assessing the crop suitability of specific crops. There is a full description in each of the six Regional Bulletins of the Soil Survey and Land Research Centre. For winter cereals the number of autumn machinery work days is particularly significant, whereas for sugar beet a combination of the number of spring (for sowing) and autumn (for harvesting) days is used.

Allocation of suitability classes for winter cereals

Machinery work days after 1 Sept		Available water (AP) Moisture deficit (MD) in mm				
		Over 40	>20 to 40	> 0 to 20	0 to –19	–20 and lower
Over 80	increasingly	Well	Well	Well	Moderate	Marginal
>50 to 80	restricted	Well	Well	Moderate	Marginal	
>20 to 50	workability	Moderate	Moderate	Moderate	Marginal	
20 and less		Marginal	Marginal	Marginal		
				> Increasing droughtiness		

Allocation of suitability classes for sugar beet

Machinery work 1 Jan–30 April–30 Dec plus 1 Sept–30 Dec		Available water (AP) Moisture deficit (MD) in mm			
		Over 40	>20 to 40	1 to 20	1 and lower
Over 120	Increasingly	Well	Moderate	Moderate	Marginal
>90 to 120	restricted	Well	Moderate	Marginal	Marginal
>60 to 90	workability	Moderate	Marginal	Marginal	
60 and less		Marginal	Marginal		
			Increasing droughtiness		

26.6.4 Suitability for grassland

A similar scheme is available for grassland suitability. Grass needs sufficient water, warmth and nutrients to grow well. Soil conditions and climate regulate the supply of moisture whereas fertilisers, notably nitrogen, are applied to sustain the supply of nutrients. The suitability of land for grass depends not only on its capacity to growth the crop but also on its ability to support animals and machinery without damage to sward or soil. Interactions between climate and soil are complex and affect the potential yield, use and husbandry of grassland in different ways.

The circumstances ideal for growth are not always those that make management, grazing or cropping easy. The large potential yields that can be achieved in a moist climate may be offset by poor ground conditions. The scheme assesses potential yield, the risk of poaching and the balance between them, to indicate relative fitness for intensive grass production.

26.7 Conclusion

The systems described above, which integrate data about climate and soil, provide a systematic and scientific method of assessing the productive capacity of a holding. Much of the requisite information can be gathered or researched from existing publications, but in some situations it may be necessary to employ a consultant to collect and interpret relevant data. By whatever means the data are assembled, they should serve to replace the vague and often arbitrary assertions about land quality that have frequently prevailed in sales particulars, valuations and arbitrations. The following are some case histories where detailed soil and climatic information played a key role in a valuation. Real names and locations have not been used.

Sale of the Denny Estate

This 400ha estate, owned by a pension fund, was put up for sale in 2006. The Agricultural Land Classification maps show the land to be equally divided between Grades 3 and 4, but the agents felt this was not a fair reflection of the land quality, since most of the estate had been converted into productive arable land in recent years. The ALC grading probably reflected in part the extent of grassland on the estate at the time of survey.

The agents commissioned a detailed soil survey of the whole estate in order to accurately reassess the ALC grading. This revealed that 85% of the estate is Grade 3, only 5% is Grade 4, and the remaining 10% is Grade 2. This information probably boosted the selling value of the estate by as much as £400 per acre.

Rent review at Lodge Farm

A tenanted farm on an estate consisted largely of sandy soils giving lower yields of sugar beet than neighbouring farms. The agent believed that this was due to poor husbandry, possibly compaction, a claim hotly denied by the farmer who was pressing for a rent reduction. A detailed soil survey and careful examination of the soil profiles under sugar beet showed no compaction, but nevertheless the roots of some plants terminated abruptly at 35cm depth. Soil analysis revealed that the colloidal content (a combination of clay, silt and organic matter) beneath the topsoil was only 1.04%, indicating that the rooting medium was almost pure sand, much more sandy than adjacent farms on the estate. Research in Denmark has demonstrated that sugar beet rooting is inhibited in sandy soils with a colloidal content of less than 5%. Thus, in this case the low sugar beet yields were due not to poor farming but to a poor farm with very sandy soils.

Dilapidations at Fenedge Farm

In 2007 the tenancy at Fenedge Farm became vacant after 12 years during which the farmer had produced carrots, potatoes and sugar beet with only an occasional cereal break. The landlord's agents were concerned that such intensive root and potato production had damaged the productive capacity of the holding through a build-up of disease, compaction and depletion of certain trace elements, for example copper and boron. A detailed soil survey and analysis revealed the following: moderate compaction in some fields but within plough depth and therefore easily removable; boron levels were adequate but copper was low in some fields, although no lower than is common in a district where regular applications of copper are required. The numbers of nematodes were moderate or low in all fields, as is typical of the district. However, infection with violet root rot was severe in many fields, and it was recommended that the production of carrots, potatoes and sugar beet should cease for about five years. Dilapidations between landlord and tenant were assessed on that basis.

Paul Wright BA, MSc

Managing Land Notified as a SSSI

27.1 Legal status of SSSI

Site of Specific Scientific Interest (SSSI) is the term used to denote an area of land notified under section 28 of the Wildlife and Countryside Act 1981 (as amended by schedule 9 of the Countryside and Rights of Way Act 2000) as being of special nature conservation interest. SSSIs comprise a nationally important series of areas of wildlife habitat, geological features and landforms and are characteristic of the natural areas of Britain. The term SSSI indicates the importance of the site and puts in place a set of procedures that require owners and occupiers to consult the relevant statutory body before carrying out specified activities that will affect the nature conservation interest of the site.

The statutory body responsible for the notification of special sites in England is Natural England (previously this responsibility was with English Nature). In Wales this role is carried out by the Countryside Council for Wales and in Scotland, Scottish Natural Heritage. The majority of the points in this chapter are relevant to all three statutory bodies, however it should be noted that it is written from an English perspective and slight differences may be found when dealing with cases in Wales and Scotland.

Under the Wildlife and Countryside Act 1981 (as amended), an SSSI is formally notified to:

- the owners and occupiers of land
- the Secretary of State for the Environment and Rural Affairs
- the local planning authority
- the appropriate water and sewage company
- internal drainage board
- the Environment Agency.

Where owners or occupiers are not known or there are any doubts about ownership, signs are placed on the site.

The documents included in a notification comprise:

- a map showing the boundaries and location
- a citation that explains the features of special interest
- a list of operations likely to damage the special interest eg changes in grazing, drainage, ploughing, etc
- a statement of the statutory body's views on how the site should be managed.

Following notification a consultation period of four months is allowed for representations and objections although the legal obligations on owners and occupiers are in force during this period. If an objection cannot be resolved by a local representative of the relevant statutory body, it is considered by a full meeting of the council of the relevant statutory body who decides whether or not the notification is confirmed.

Within nine months of notifying a site, the statutory body must confirm or withdraw a notification. A modification of the notification may be carried out to reflect an objection.

Once the Secretary of State, the local planning authority and the owners and occupiers have been sent the notification the site becomes an SSSI and is registered as a local land charge. When a site subsequently changes ownership, its SSSI status is discovered when a search is made with the Land Registry. Following the Countryside and Rights of Way Act (CROW) in 2000 an owner of an SSSI has a legal obligation to notify Natural England within 28 days of a change in ownership or occupation.

27.2 Carrying out management on land notified as an SSSI

If an owner or occupier wishes to carry out an operation listed in the SSSI notification as being likely to damage the site, she must give notice in writing of the intention. If an operation is carried out without the consent of the statutory body it is considered to be an offence unless there is a reasonable excuse, with a fine of up to £20,000 in the magistrates court or an unlimited fine in the Crown court.

Following written notice of an operation the statutory body will respond giving consent, refusal or an invitation to the owner or occupier to discuss modifying the operation. Management objectives for the site will be discussed and a management agreement may be offered at this stage to ensure positive management. If the statutory body does not respond within four months consent is deemed to be refused, although there is a right to appeal within two months to the Secretary of State. Prior to the CROW Act in 2000, if the statutory body did not respond within four months the owner or occupier was able to proceed with the operation. This major change in the legislation reflects the enhanced protection given to SSSIs by CROW.

The list of operations described as being likely to damage in a notification may often appear daunting. This is because legally all operations which could conceivably damage the special interest must be listed. This applies to operations applicable to present land use and any future land use change, as the list cannot be added to at a later date without undergoing complete re-notification. The list also covers the whole site, although some of these operations may apply only to certain parts.

The list of operations requiring consent is essentially a mechanism for consultation between the owner or occupier and the statutory conservation body on the management of the site. Consent may be given for a proposed operation, it may be refused or consultation may result in a mutually acceptable modification, for example altering the timing of agricultural operations.

To avoid neglect of SSSIs, the CROW Act 2000 introduced a new set of tools for the statutory body to use, to ensure the required management is delivered to maintain the special interest. A management scheme may be served where the site has deteriorated through neglect or where it is not possible to secure future management through an appropriate agri-environment, forestry or section 15 or section 7 management agreement. If an owner or occupier does not implement the management scheme a management notice may be served requiring the SSSI to be managed in accordance with the management scheme.

The management scheme and notice are only used if the statutory body cannot agree appropriate management with the owner or occupier. In the majority of cases advice is given by the relevant statutory body on the management needed to sustain and enhance the special interest. Information is provided on environmental land management schemes available to support and deliver the desired management (in England the Environmental Stewardship Scheme and for woodland sites the England Woodland Grant Scheme). Where it is not possible to bring a site into positive management through advice or other schemes, the statutory body may offer a section 15 or section 7 management agreement.

A number of SSSIs have an additional European designation with additional protection in place through the Conservation (Natural Habitats &c) Regulations 1994. Further information is given in 27.9.

27.3 Management agreements on SSSIs

A management agreement is an agreement made between a relevant authority and any person with an interest in land for the purpose of managing that land to conserve or enhance its natural beauty or amenity or to promote its enjoyment by the public. There are a number of different types of management agreements, some of which can be entered into with local authorities and voluntary conservation bodies such as local conservation trusts. SSSI land is normally entered into a management agreement under the Environmental Stewardship Scheme or England Woodland Grant Scheme however the statutory body is also able to enter into agreements under section 15 of the Countryside Act 1968 as amended or section 7 of the Natural Environment and Rural Communities Act 2006.

Entering into a section 15 or section 7 management agreement on an SSSI will provide the financial support to enable the required management to be undertaken and depending on the management required to maintain the conservation interest, this may cover all or part of the cost. Agreements may be offered under Natural England's Wildlife Enhancement Scheme although this scheme ended at the end of 2007 with a successor scheme awaiting EU approval. The approach and level of payments under section 15 agreements have varied considerably as changes to the underlying legislation have been made.

The major change in approach came with the introduction of the CROW Act in 2000. This act required payments to be consistent with European Union State Aid Guidelines. The guidelines require agreements to comply with the Rural Development Regulation (RDR) and implementing regulation. Article 24 of the RDR gives the basis of calculating payments as income foregone, additional costs arising from the commitment given and the need to provide an incentive. Using this methodology the statutory body produces a set of standard payments for management of different habitat types and a contribution to the cost of management works such as fencing or scrub control.

27.4 Professional costs relating to section 15 and section 7 agreements

The Guidelines on Management Agreement Payments and Other Related Matters published in 2001 confirmed the principles for reimbursement of professional fees. On completion of a management agreement under section 15 of the Countryside Act (as amended), reasonable professional fees may be reimbursed by the statutory body. Fees are not reimbursed in relation to professional advice concerning the notification of an SSSI, responding to a management scheme or notice, entering an environmental stewardship scheme or for other estate management issues. As agreements are now standard it is considered reasonable for the statutory body to agree a ceiling on fees at the beginning of the negotiations. Agents should charge clients for the fee due and the statutory body will reimburse the fee net of VAT at the agreed rate.

27.5 Environmental Stewardship Scheme

In England the main scheme which delivers management on SSSIs is the Environmental Stewardship Scheme. Similar schemes exist in Wales (Tyr Gofal) and Scotland (Natural Care) although the options available and payment rates differ.

The Environmental Stewardship Scheme has three elements: Entry Level Stewardship (ELS), Organic Entry Level Stewardship (OELS) and Higher Level Stewardship (HLS). The ELS and OELS schemes are whole farm schemes and acceptance is guaranteed provided a land manager meets the scheme requirements. The HLS is a discretionary and competitive scheme and can be combined with ELS or OELS to deliver a wide range of environmental benefits in high priority areas and situations. The primary objectives of the Environmental Stewardship Scheme are to conserve wildlife and biodiversity, maintain and enhance landscape quality and character, provide natural resource protection, protect the historic environment and promote public access and understanding of the countryside.

As SSSIs are nationally important areas of wildlife habitat and key biodiversity indicators they are considered to be a high priority for the scheme. If an HLS application includes an SSSI and the options chosen will deliver the required management this will increase the likelihood of successful application.

ELS and OELS agreements for the last five years with a simple application process and payments per hectare of £30 for ELS and £60 for OELS. The payment rates reflect the points worth of management options selected by the applicant with organically managed land attracting an additional 30 points and therefore an additional £30 per ha. Under both ELS and OELS there are 50 simple management options to choose from to deliver the required 30 points per ha.

Under HLS the agreements are for 10 years and payments received reflect the scheme options chosen by the applicant. There is an option for either party to withdraw from the agreement after five years without any penalty as long as the required notice is given.

The applicant or agent is required to consult with Natural England prior to application to discuss the suitability of the application. Following discussion and in some instances a visit by the Natural England advisor, the applicant is advised as to whether the application is considered to be of high priority and good value for money an whether it should proceed. The applicant or agent will then need to prepare a farm environmental plan (FEP) to identify the most important environmental features on the holding, their condition and provide a guide to the most appropriate options. Payments for FEPs range from £395 to £3,350 and relate to the area of land included. If an application is submitted without confirmation from the Natural England advisor that the application should proceed and the application is rejected, there will be no payment made for the FEP.

For holdings that include SSSI land there is a requirement during the FEP process to discuss the required management with the Natural England advisor to ensure it will deliver favourable condition. If the application is successful, consent for the management operations included in the agreement will be given. Full details of the application process and terms and conditions of the scheme are provided in the scheme handbooks.

Annex One gives an example of standard payments, both annual management payments and capital works payments under the HLS (as at October 2007).

27.6 England Woodland Grant Scheme

The England Woodland Grant Scheme consists of six grants for the stewardship and creation of woodlands and is operated by the Forestry Commission. Funding is managed on a regional basis with grants focused to meet the priorities laid out in the Regional Forestry Framework action plans. Applications received on SSSI woodlands will be discussed between the Forestry Commission and Natural England to ensure agreements deliver favourable condition and the necessary consents are obtained. The Woodland Improvement Grant supports capital projects to sustain and increase public benefit in woodlands with one of the four national funds available only on SSSIs. An 80% contribution to works is available where the work will address the key threats and recover condition on sites that are currently considered unfavourable.

27.7 Inheritance tax

An owner of land which is of outstanding scenic, scientific or historic interest may apply to the Treasury to obtain exemption from future inheritance tax (IHT) liability (section 30 of the Capital Transfer Tax Act 1984 (as amended)). If the Treasury accepts the application, the owner will be required to give assurances regarding the future protection of the land, similar to those included in a management agreement. It should be noted that it is not possible to receive both IHT exemption and payments under an agreement for the same restrictions. In some instances additional management to enhance the wildlife interest of the site may be funded by the statutory body through a management agreement in addition to the IHT exemption.

27.8 Capital values of SSSI land

There has been much speculation about the impact of SSSI designation on the capital value of land and a number of research projects have been commissioned in this area. As all valuers know, land prices fluctuate because of many complex reasons and values are not always linked simply to the productive capacity of the land. SSSI land as other land varies significantly in both accessibility and economic potential and the objectives of buyers vary. Some seek land for commercial enterprise and are therefore not interested in land with restrictions while others are prepared to pay a premium for land with a special nature conservation interest. The increased availability of management agreements on SSSI land with attractive payment rates when compared with many farm enterprises is also considered to have the effect of stabilising farm returns. This stability is of benefit to risk adverse farmers and makes land with SSSI designation more attractive.

The studies to date have concluded that statistically there is no evidence to suggest that in general terms SSSI designation has a significant effect on land prices although both positive and negative fluctuations have been noted on individual land sales.

27.9 Additional provisions for SSSIs with European designations

The Conservation (Natural Habitats &c) Regulations 1994 came into force in Europe in October 1994. The Secretary of State compiled a list of sites of European importance and a number of existing SSSIs have now become Special Areas of Conservation (SACs) or Special Protection Areas (SPAs). These comprise the Natura 2000 network of sites, which have been selected as best examples of a prioritised list of particular habitats and species. The designation of sites in England was confirmed by the Secretary of State on 1 April 2005.

The designation requires competent authorities (defined as any minister, government department or public body of any description) to take into consideration the EC Habitats Directive that requires the favourable conservation status of natural habitats and species to be maintained or restored. The directive places a general duty on member states to avoid habitat deterioration and significant species disturbance, and to ensure safeguards are applied whenever there is a proposal that potentially threatens a European site.

The sites with the additional designation are all SSSIs and the same consultation and consent procedures apply. The main differences apply if a project (for example a project needing planning permission or Environment Agency consent) is proposed on or adjacent to a Natura 2000 site. If the project would not adverse affect the integrity of a site the competent authority may approve it.

If following an environmental assessment it is concluded that the project would damage the wildlife interest and a management agreement is not agreed with the statutory body, the Secretary of State may make a Special Nature Conservation Order (regulations 22–27, 91–93 and schedule 1) to prevent the operation being carried out. The order applies indefinitely and if an owner is able to prove the value of the land is reduced by the implementation of the order he is entitled to claim compensation. Compensation is assessed in accordance with the provisions of the Land Compensation Acts 1961 and 1973.

The competent authority may approve a project if there are 'imperative reasons of overriding public interest, including those of a social or economic nature'. Approval for development on high priority sites may only be given for 'reasons relating to human health or public safety or if they have beneficial consequences of primary importance to the environment'.

The regulations also require competent authorities to review outstanding decisions, permissions, consents and other authorisations, not yet completed, and revoke or amend them if necessary. Under existing legislation landowners and public bodies have permitted development rights for some activities. On SSSIs if the activity is listed as a potentially damaging operation, the statutory body would need to give approval. The regulations require that any permitted development likely to significantly affect an SAC or SPA will need planning permission. Compensation is payable by the planning authority if an existing planning permission is modified or revoked or permitted development is prevented.

Glenys Tucker BSc (Hons), MRICS

Annex One

Examples of annual management payments and capital works payments available under the Higher Level Scheme (October 2007)

Payments

Grassland options

	£ per ha
Maintenance or restoration of species rich, semi-natural grassland	200
Maintenance of wet grassland for breeding waders	335
Maintenance of wet grassland for wintering wades and wildfowl	255
Maintenance of semi-improved or rough grassland for target species	130
Supplement for hay making	75

Resource protection options

Arable reversion to unfertilised grassland to prevent erosion or run-off	280
Infield grass areas to prevent erosion or run-off	350
Seasonal livestock removal on grassland with no input restriction	40

Moorland and rough grazing options

Maintenance of moorland	40
Maintenance of rough grazing for birds	80
Shepherding supplement	5
Seasonal livestock exclusion supplement	10

Capital items

Hedgerow restoration or planting	£5 per m
Sheep fencing	£1.80 per m
Deer fencing	£4 per m
Coppicing bankside trees	£29 each
Water supply	£2 per m
Water trough	£85 each

Agricultural Tenancies Act 1995

This Act came into force on 1 September 1995 and applies to farm business tenancies granted under it. Apart from arranging lettings, preparing agreements and management, valuers are mainly concerned with claims for fixtures, improvements and rent reviews.

Compensation on termination of farm business tenancy (Part III)

28.1 Fixtures

Section 8 of the Act gives the tenant a right to remove:

(a) any fixtures (of whatever description) affixed, whether for agriculture or not, to the holding by the tenant under a farm business tenancy and
(b) any building erected by him on the holding.

These fixtures can be removed by the tenant at any time during the tenancy, or at any time after the termination of the tenancy when he remains in possession as tenant (whether under a new tenancy or not).

However, this right shall not apply:

(a) to a fixture affixed or a building erected in pursuance of some obligation

(b) to a fixture provided or building erected instead of some fixture or building belonging to the landlord

(c) to a fixture or building in respect of which the tenant has obtained compensation under section 16 of the Act or otherwise

(d) to a fixture or building in respect of which the landlord has given his consent under section 17 of the Act on condition that the tenant agrees not to remove it and which the tenant has agreed not to remove.

This section further stipulates that in removal the tenant shall not do any avoidable damage to the holding and shall make good any damage occasioned by the removal. It further provides that the section applies to a fixture or building acquired by the tenant equally as it applies to a fixture affixed or building erected by him.

From a tenant's removal point, the determination of what are fixtures and their ownership is most important and should, if at all possible, be agreed with the landlord prior to removal. In most cases matters can be agreed, as, if the fixture is of value to the holding, the landlord will normally wish to purchase unless, for instance, there is going to be a change of farming policy whereby the fixture may be of no value in the future use of the holding.

28.2 Tenant's improvements

Generally speaking, subject to the conditions mentioned below, tenants holding under a farm business tenancy are entitled to compensation for tenant's improvements.

Section 15 of the Act, gives the meaning of tenant's improvements under a farm business tenancy as being:

(a) any physical improvement which is made on the holding by the tenant by his own effort or wholly or partly at his expense, or

(b) any intangible advantage which —
  (i)   is obtained for the holding by the tenant by his own effort or wholly or partly at his own expense and
  (ii) becomes attached to the holding.

Note the distinction between routine improvements and other improvements: see paragraph 28.2.8 below.

28.2.1 Right to compensation

Section 16 of the Act states that on the termination of the tenancy, on quitting the holding the tenant is entitled to obtain from the landlord compensation in respect of any tenant's improvement.

However, the tenant is not entitled to compensation under this section in respect of:

(a) any physical improvement which is removed from the holding or
(b) any intangible advantage which does not remain attached to the holding.

This section further stipulates that section 13 and schedule 1 to the Agriculture Act 1986 (compensation to outgoing tenants for milk quota) shall not apply in relation to a farm business tenancy. Milk quota is probably an intangible advantage.

28.2.2 Consent of landlord as condition for tenant's improvements

Section 17 of the Act states that a tenant shall not be entitled to compensation under section 16 of the Act in respect of tenant's improvement *unless the landlord has given his written consent to the improvement.*

This consent may be given in the tenancy agreement (eg requirements of good husbandry such as liming, applying farmyard manure, applying fertilisers, sowing young seeds, requirement to plant certain crops, etc) or separately. This consent may be unconditional or on condition that the tenant agrees to a specified variation in the terms of the tenancy (which variation must be related to the tenant's improvements in question). Section 17 does not apply in any case where the tenant's improvement consists of planning permission.

28.2.3 Conditions in relation to compensation for planning permission

Section 18 of the Act stipulates the tenant shall not be entitled to compensation under section 16 of the Act in respect of tenant's improvement which consists of planning permission unless:

(a) the landlord has given his written consent to the making of the planning application

(b) that consent is expressly to be given for the purpose of

    (i) enabling a specified physical improvement falling within paragraph (a) of section 15 of the Act lawfully to be provided by the tenant or

    (ii) of enabling the tenant lawfully to effect a specified change of use and

(c) on the termination of the tenancy, the specified improvement has not been completed or change of use effected.

Any consent given to the application may be unconditional or on condition that the tenant agrees to a specified variation in the terms of the tenancy (which variation must be related to the physical improvement or change of use in question).

28.2.4 Amount of compensation for tenant's improvement

Section 20 of the Act states that the amount of compensation payable to the tenant in respect of tenant's improvement shall be 'an amount equal to the increase attributable to the improvement in the value of the holding' at the termination of the tenancy as land comprised in a tenancy.

Where the landlord and tenant have entered into an agreement whereby any benefit is given or allowed to the tenant in consideration of the provision of a tenant's improvement, the amount of compensation shall be reduced by the proportion which the value of the benefit bears to the amount of the total cost of the improvement (eg a reduction should be made where the landlord provides materials and the tenant carries out the work such as land drainage). However, in such cases it is important that where the improvement is effected, detailed records of the costs involved by either party should be correctly apportioned and recorded.

Where a grant is made to the tenant out of public funds for the improvement, the compensation must be proportionally reduced.

Where an improvement has been completed or change of use (by planning permission) effected, any value attributable can be taken into account.

Section 20 does not apply where the tenant's improvements consist of planning permission.

28.2.5 *Amounts of compensation for planning permission*

Section 21 states that the compensation payable to the tenant under section 16 of the Act in respect of a tenant's improvement which consists of planning permission shall be 'an amount equal to the increase attributable to the fact that the relevant development is authorised by planning permission in the value of the holding at the termination of the tenancy as land comprised in a tenancy'.

Where the landlord and tenant have entered into a written agreement where benefit is given or allowed to the tenant in consideration of obtaining planning permission by the tenant, the amount of compensation shall be reduced by the proportion which the value of the benefit bears to the amount of the total cost of obtaining permission.

26.2.6 *2006 Regulatory Reform Order (TRIG)*

This order limits the amount of compensation payable under section 16 of the 1995 Act, where the parties have agreed in writing to the lesser of

(a) the amount determined in accordance with subsections (1)–(4) of section 20 of the 1995 Act and
(b) the compensation limit.

The compensation limit is the amount agreed in writing or where they cannot agree to the cost of the improvement. This excludes planning permission.

28.2.7 *Settlement of claims for compensation*

Under section 22 of the Act, no claim for compensation shall be enforceable unless 'before the end of the period of two months beginning with the date of the termination of the tenancy the tenant has given notice in writing of his intention to make the claim and of the nature of the claim'.

Claims are normally settled by negotiation and agreement and if not settled, and an arbitrator has not been appointed by agreement made following the section 12 notice, either party may, after the end of four months beginning with the termination of the tenancy, apply to the President of the RICS for the appointment of an arbitrator.

If such application for the appointment of an arbitrator relates wholly or partly to compensation in respect of a routine improvement (see below) which the tenant has made, and an application can be made at the same time seeking consent to the provision of the improvement, the President shall appoint the same arbitrator to deal with both applications.

Arbitration is under the provisions of the Arbitration Act 1996.

28.2.8 Successive tenancies

Section 23 of the Act states that when a tenant under a farm business tenancy has remained on the holding during two or more such tenancies, he shall not be deprived of his right to compensation under section 16 of the Act if the improvement was effected during a tenancy other than the one at the termination of which he is claiming (unless the parties have agreed otherwise).

28.2.9 Reference to arbitration

Under the provisions of section 19 of the Act where the landlord refuses to give consent, and fails within two months to give consent following a written request by the tenant for such consent or the landlord requires variation in the terms of the tenancy as a condition of giving such consent for a tenant's improvement, the tenant may on written notice to the landlord demand that the question shall be referred to arbitration. This notice may not be given in relation to any tenant's improvement which the tenant has already provided or begun to provide, unless that improvement is a routine improvement.

The procedure for an arbitration reference is detailed in section 19 and must be carefully followed. Note the difference between routine improvements and other improvements.

Routine improvements are a class of tenant's improvements whose eligibility for compensation at the end of the tenancy can be referred retrospectively to arbitration. Section 19(10) of the Act defines these tenant's improvements as:

(a) any physical improvement made in the normal course of farming the holding or any part of the holding and

(b) does not consist of fixed equipment or an improvement to fixed equipment but excludes any improvement whose provision is prohibited by the terms of the tenancy.

Fixed equipment includes any building or structure affixed to land and to any works constructed on, in, over or under land and also includes anything grown on land for a purpose other than use after severance from land, consumption of the thing grown or its produce, or amenity.

Examples of routine improvements are for instance some items of tenant right such as growing crops, pastures laid down at the expense of the tenant, cultivations and improvement such as applying lime and fertilisers.

Examples of items that will not be routine improvements include quotas, abstraction licences, caravan site licences, planning permission, discharge consents, fixed equipment such as buildings, fencing, permanent drainage, piped water supplies, installation and wiring of electricity, hop wire work, protection of fruit trees, produce, consumable stores and severed crops.

28.2.10 Practical points arising on the compensation provisions of the Act (Part III sections 16–27)

There are many anomalies and lack of clear definitions which are likely to give rise to future disputes on the termination of a tenancy.

It is therefore strongly recommended:

(a) that tenancy agreements, especially agreements of over two years' duration should be prepared with utmost care. For instance, consideration should be given to the question of severed crops on harvesting, straw and silage which are tenant's chattels and which he ordinarily can remove. Provision should be made in the agreement if these are to be left on the holding and the basis of valuation.

(b) prior to the termination of a tenancy (preferably a year in advance) the parties should agree if possible on what is to be left on the farm, ie silage, hay, straw, growing roots, tenant's pastures, compensation for unexhausted value of lime, phosphate, potash and farmyard manure, slurry, etc. Also if possible to agree the value of other tenant's improvements fixtures, etc.

Such agreement will not override the compensation provisions of Part III of the Act (and arbitration) — but at least it is hoped that if an agreement is made it will be honoured by the parties. Section 26 does not allow contracting out.

It is further observed that the statutory basis of compensation (section 20) is the amount equal to the increase attributable to the improvement in the value of the holding at the termination of the tenancy as land comprised in a tenancy. Section 21 defines the compensation payable in respect of obtaining planning permission.

The measure of compensation in practical terms is by means of assessing the increase in the value of the holding by taking the difference in rental values of the holding, with and without the improvement and then capitalising the difference by an appropriate year's purchase. If a rent has not been reviewed for some time, the current rental value should be considered and not what is paid. The year's purchase used should bear relation to the expected life of the improvement and also on any market evidence. of interest rates. It is considered that the increase in value due to the improvement should be calculated on the basis that the holding is to be relet on a farm business tenancy.

The valuation should not be too difficult in the case of major and expensive improvements, such as the erection of a large building. However, it will be a different matter in the case of routine improvements such as normal acts of husbandry, eg liming, applying fertilisers, tenant's pastures, growing crops, etc. Indeed it is observed that these compensation provisions for routine improvements are most unsatisfactory and retrograde on compensation provisions for lettings under the Agriculture Holdings Act 1986.

It is observed that the various statutory instruments under which compensation is calculated in the Agricultural Holdings Act 1986 (SI 1978 No 809, SI 1981 No 822, SI 1983 No 1475) have no statutory force under Agricultural Tenancies Act 1995 tenancies and cannot be used, unless by agreement, prior to the termination of the tenancy, the parties agree to adopt them and even then, if there is a dispute, they are not enforceable unless they are matters not covered by the definition of improvements in the 1995 Act (eg certain tenant right matters).

It is further emphasised that 'writing down' over a period of years of the cost of an improvement, eg a corn store costing £25,000 over 25 years, is no longer enforceable as a condition of landlord's consent.

These somewhat ambiguous compensation provisions, particularly in respect of routine improvements, are likely to give rise to difficulties in interpretation, assessment of compensation and resolution unless there is a measure of prior agreement between the parties (and such agreement cannot be enforced in the event of a dispute — but rather only the legal measure as provided in Part III of the Act). The CAAV is recognising these difficulties and has issued draft guidelines on valuation which include the following items.

- Acts of husbandry — the reasonable cost of operations based on CAAV costings.
- Mole drainage — the reasonable cost of the work depreciated over a three- to six-year period dependent on the quality and efficiency of the work and soil type.
- Sub soiling — reasonable cost of the work based on CAAV costings written off over a period of years, unlikely to be more than three. Subsoiling tramlines is not an improvement, but maintenance.
- Growing crops — reasonable cost of seeds, cultivations (CAAV costings) fertiliser, sprays. When the crop is mature it should be valued as a standing crop less cost of harvesting and allowing a risk factor.
- Pastures and permanent crops — (eg fruit trees, asparagus beds, rhubarb beds and crops grown for Bio-mass.) The flow of future benefit or income should be valued. (Again, not easy to assess.)
- Liming of land — in the first full growing season after application, the value will be cost. Thereafter it should be depreciated by equal annual amounts for four to eight growing seasons from application date, dependent on nature of lime and soil types.
- Application of purchased inorganic manure and fertiliser — where no crop taken cost of fertiliser delivery and application. Thereafter: nitrogen (N), potash (K_2O) and phosphate (P_2O_5): nitrogen after one crop taken — nil; potash after one growing season — half cost, after two growing seasons quarter cost, after three growing seasons nil. It is to be reduced where applied to potato, leafy brassicas and pastures. Phosphate — depreciated after one crop taken over the following growing three seasons. (It is recommended that a soil analysis should be made.)
- Application of farmyard manure — this should be assessed on the cost of the work and market value of the manure. (Note FYM in store and slurry in store is likely to be a chattel rather than an improvement.)

28.3 Rent review

The rent review provisions of the Act, section 9 apply unless the tenancy agreement expressly states that the rent is not to be reviewed during the tenancy or provides that the rent is to be varied, at specified times by a specified amount or in accordance with a specified formula which does not preclude a reduction and which does not require or permit the exercise by any person of any judgment or discretion in relation to the determination of the rent of the holding, but otherwise to remain fixed.

Notices, requiring arbitration on the rent, unless the parties have agreed otherwise in writing, in the tenancy agreement, must be served in accordance with the requirements of section 10 of the Act.

Under section 13 of the Act, in pursuance of a statutory review notice, the arbitrator shall determine the rent properly payable at the review date and accordingly shall, with effect from that date increase or reduce the rent previously payable or direct it shall continue unchanged.

The rent properly payable is the rent at which the holding might reasonably be expected to let in the open market by a willing landlord to a willing tenant taking into account all relevant factors, including in every case, the terms of the tenancy. The arbitrator shall disregard any increase due to tenant's improvement other than those the tenant is under an obligation to provide in his tenancy agreement, or where allowance or benefit has been given by the landlord in consideration of its provision and where the tenant has received compensation from the landlord in respect of it.

The arbitrator is to disregard any effect on the rent that the tenant is in occupation of the holding and must not fix a rent at a lower amount by reason of dilapidation, deterioration or damage to the property caused or permitted by the tenant.

Under section 12 of the Act, if an arbitrator has not been appointed by agreement, either party may apply, at any time during the period of six months ending with the review date, to the President of the RICS to appoint an arbitrator.

It is important to note that under section 10 of the Act, to trigger a rent review either party must give a written notice requiring arbitration to review the rent. Unless agreed in writing that the rent is to be reviewed from a specified date or dates or intervals, the review date must be a date as from which the rent could be varied in the agreement (unless it expressly states that the rent is not to be reviewed during the tenancy).

The review date must be at least 12 months but less than 24 months after the day on which the statutory review notice is given.

The rent arbitration will be under the provisions of the Arbitration Act 1996 but the Regulatory Reform Order 2006 allows review by an independent expert provided that the agreement is after 19 October 2006, but does not provide for upward review.

Example notice of claim compensation for tenant's improvement under the Agriculture Tenancies Act 1995

AGRICULTURE TENANCIES ACT 1995
NOTICE PURSUANT TO SECTIONS 16 & 22

To: Mr JM Smythe-Andrews (Landlord)
        Horton Hall
        Exterton EX1 5XS
Re: The holding known as Starvecrowe Hill Farm, Exterton, EX1 5UR

I, Mr J Willes, Starvecrowe Hill Farm, Exterton EX1 5UR, HEREBY GIVE YOU NOTICE that I intend to claim compensation from you under section 16 of the above Act in respect of improvement for which you gave consent in writing on the dates given below. Full details of the improvements, nature and amount of claim are set out in the Schedule below.

SCHEDULE
1. **Fixed equipment**
    1.1 Erection of steel framed, asbestos and profile sheeted
        Dutch Barn 90' × 40' (27.43m × 12.19m) in OS No 1234.
        Written consent given in Messrs Haile and Well, your Land
        Agents in letter 1 November 2005

        Claim — Rental value of holding without barn
        = £10,000

        Rental value of holding with barn = £11,000
        Increased rental value = £1,000 × 12 yp   £12,000.00

    1.2 Plastic tube drainage of OS No 4824.10 acres
        (4.04ha) with back fill. Written consent from
        Messrs Haile and Well, 1 November 2005

        Claim — Rental value of holding without drainage
        = £10,000

Rental value of holding with drainage = £10,500
Increased rental value = £500 × 10 yp   £5,000.00

1.3 Planning Consent to Change Use of OS No 5678
5.2 acres (2.10ha) into permanent Caravan Site
Written consent from Messrs Haile and Well,
1 November 2005

Claim — Rental value without consent
= £10,000
Rental value with consent
= £12,000
Increased rental value = 2,000 × 10 yp   £20,000.00

2. Routine improvements

Written consent for these were given by Messrs Haile and Well, 10 January 2007 and basis of claim agreed in their letter of 1 July 2007

2.1 OS No 1018.15.85 acres (6.41ha). Rustler Winter
Wheat Growing crop of winter wheat at cost of
cultivations, seeds, fertilisers, sprays and
enhancement value,
15.5 acres (6.41ha) @ £82.50 per acre (£203.85/ha)   £1,278.75

2.2 OS No 5861.10.30 acres (4.16ha). Lagune Winter
Barley Growing crop of winter barley at cost of
cultivations, seeds, fertilisers and enhancement value,
10.3 acres (4.16ha) @ £74.80 per acre (£184.83/ha)   £770.44

2.3 OS No 4440.18.86 acres (7.43ha), 5 year Ley
Sown after winter barley on 1 September 2007.
Not grazed. Cost of cultivation, seed, fertiliser and
enhancement value
18.5 acres (7.63ha) @ £101.50 per acre (£250.80/ha)   £1,877.75

2.4 OS No 6861. 20.62 acres (18.34ha), subsoiled and
ploughed Subsoiling and ploughing
20.5 acres (18.34ha) @ £27 per acre (66.71/ha)   £553.50

2.5 Application of Fertiliser
March 2005
20 tonnes Potash, ex Messrs Jakes and Co
Cost £2,100, applied to cereals fields
Claim half of cost of £2,100   £1,050.00

March 2006
10 tonnes Phosphate ex Messrs Jakes and Co
Cost £1,100 applied to grassland
Claim three-quarters of cost of £1,100   £825.00

2.6   Application of Lime
April 2006
40 tonnes ground limestone ex Rock Fare Ltd
cost £541 applied Claim £541 depreciated over
six years, 4/6 × £541   £360.66

E&OE
Dated this 3rd day of February 2008

Signed . . . .
Williams, Davies, Rees and Thomas
Chartered Surveyors
Dolanog
Welshpool, Powys

Authorised Agents for and on behalf of the Tenant, Mr J Willes

Valuations in Court Proceedings

29.1 Introduction

This chapter will focus on valuations provided in the course of litigation proceedings and will deliberately not refer to valuations provided in an independent expert capacity, which are dealt with elsewhere in this book. An individual providing a valuation for use in court proceedings is referred to as an expert witness. Furthermore, this chapter deals only with expert witness valuations carried out for courts in England and Wales, governed as they are by the English legal system, different to that of Scotland and Northern Ireland where other rules may apply. Indeed, much of the guidance referred to in this chapter has been specifically drawn up under the English legal system.

29.2 What is an expert witness?

There are several definitions of the term expert witness. The *Oxford English Dictionary* defines the word expert as 'a person who is very knowledgeable about or skilful in a particular area'. However, one definition from the American courts which concisely sums up an expert witness is 'a witness who has been shown to the Court to be qualified by their special knowledge skill or experience (scientific, technical, or other) and who can testify as an expert in a specific field. Expert witnesses can give opinions based on their special knowledge or skill.

29.2.1 Courts presiding over disputes will often issue instructions for the parties in the dispute to appoint an expert witness to provide assistance of a technical nature where the court decides it is necessary. Disputes over valuation are typical of this reliance by the judicial system on outside help. The expert witness is unique in the court room environment in that he is the only person entitled to give his opinion, referred to as giving opinion evidence.

29.3 Requirements of the expert witness

As an expert witness, the valuer will have to adhere to a collection of rules, protocols, practice statements and guidance notes in producing his report. Moreover, he will need to be sufficiently familiar with their content and the rapidly increasing body of relevant case law applicable to expert witnesses to be able to undertake his instructions with confidence. More of this later.

29.3.1 The Civil Procedure Rules

In a civil court the expert witness is governed in his approach by the Civil Procedure Rules (CPR), most particularly part 35, issued by the Ministry of Justice. The CPR were brought about following the review of the procedures adopted in pursuing or defending a case through the civil courts by the Woolf Reforms brought into effect on 26 April 1999. These rules are a procedural code with the overriding objective of enabling the court to deal with cases justly. Dealing with a case justly includes, among many others, ensuring a consistent and objective approach by expert witnesses. Prior to the reforms carried out by Lord Woolf at the end of the 1990s, the rules governing the duties of an expert witness were by no means as clear, and much essential clarification existed within individual court judgments. The clearest summary of the guidance was set out by Cresswell J in the case of the *Ikarian Reefer* (1993). Valuers taking on their first instruction from the court would do well to consider the guidance from this case as an important additional measure by which to approach their role. Part 35 addresses the expert's overriding duty, which is to the court, rather than to either party in the case, regardless of who has instructed him or may be paying his fees. Instructing parties are also governed by part 35 in the procedure they should adopt, and extends to stipulate what they may or may not ask an expert to do and what to expect of an expert. Part 35 is essentially a template for experts and those instructing them in how that expert evidence may be sought and presented.

29.3.2 The Civil Justice Council Protocol

The Civil Justice Council Protocol for the Instruction of Experts to give Evidence in the Civil Courts, to give it its full title, which came into effect in 2005, assists in the interpretation of the CPR part 35 and its associated Practice Direction. Those accepting instructions to act as an expert witness are required not only to be experts in their field, but also to be familiar with CPR part 35, its Practice Direction and the Civil Justice Council Protocol. Indeed, the expert held by the court as failing to adhere to this guidance may have costs awarded against him personally (*Phillips* v *Symes* 2004).

29.3.3 The RICS Practice Statement and Guidance Note Surveyors Acting as Expert Witnesses

At the time of writing, the second edition of this publication was under review. Unsurprisingly, as indicated by the title, this publication is made up of two parts: a practice statement and associated guidance. The practice statement identifies such items as the surveyor or valuer acting in an expert witness capacity must follow, although there is scope in exceptional circumstances to deviate from it. These include the expert's duty to the court, correct procedure if the instructions are changed after the initial appointment, inspecting the property, conducting meetings with other experts, including how to agree facts and resolve differences, amending your opinion, and how to conduct yourself in the event you are appointed jointly by both parties.

The guidance note provides 'further material and information on good practice considered to be generally appropriate where a surveyor is required to give expert evidence'. This includes interpretation of the difference between duty to a client and duty to the court, how to deal with fees in various circumstances such as when booked to attend a court hearing, which may be cancelled, and so on. A sample terms of engagement is included as an appendix to the guidance note.

29.3.4 The RICS Appraisal and Valuation Standards ('The Red Book')

Chartered surveyors are bound by the requirements of the Red Book in carrying out valuations, except in a small number of cases. Advice given in the course of litigation is one such occasions: 'Valuations prepared in anticipation of giving evidence as an expert witness before a court, tribunal or committee in connection with litigation related to matters in which property value is in dispute. However, adopting the principles and definitions in any Standards that are relevant and practical may give evidence credibility and help withstand cross-examination'. Despite this express guidance, many valuers, and ironically also the RICS in its practice statement and guidance note *Surveyors Acting as Expert Witnesses*, also choose to follow the guidance of the Red Book as best practice in preparing a valuation report. However, one should not follow the Red Book headings slavishly where they may not be appropriate in the circumstances. The valuer will be expected to exercise sound judgment in interpreting best practice.

29.4 Appointments under the CPR

An expert may be appointed by either one of or both parties in a dispute. The CPR actively encourages the appointment of a single joint expert, one who is jointly instructed by both parties, on the grounds that one experts view may be more expedient to the court in resolving the issues in dispute than two or more. However, the CPR doesn't rule out the possibility that each party may have justifiable cause to appoint its own expert and so provides guidance for experts in how to conduct discussions designed to eliminate issues where both agree and address areas where the experts' opinion may differ.

29.4.1 Content and form of a report

An example report is provided at the end of this chapter, but the salient elements of what the report should contain are spelt out for clarity. The report must be addressed to the court and not to the party from whom the expert received his instructions. Evidence must be presented in an organised and logically referenced way, with brevity, distinguishing where possible, between matters of plain fact, expert observations and expert inferences. In preparing any written evidence, the expert must consider all matters relevant to the instruction.

As regards layout, the front sheet should reveal, not obscure, the name of the expert witness, the proceedings and the nature of the evidence, the subject of the report and the date it was produced. Thereafter, the report often takes the following form:

- introductory material
- a brief resume of qualifications and relevant professional experience
- chronology as to the expert's involvement in the case
- the factual or assumed background of the case, and
- the issues which the expert proposes to address in the report
- identify the expert's conclusions and give reasons
- provide a statement that the expert understands his duty to the court and has complied with that duty.

9.4.2 Declarations

Most instructing solicitors will ask the valuer to confirm his experience and eligibility to carry out an expert witness valuation and will expect him to make an appropriate declaration within his report that he is suitably qualified and experienced to provide the opinion sought by the court. The report will also need to include a declaration to the effect that the expert witness understands that his overriding duty is to the court, a full example of which is included in the worked example at the end of this chapter.

20.5 Contingency fees

The issue of contingency fees is controversial. This method of charging fees is common for valuers acting as negotiators. The problem arises when the role changes to that of expert witness, effectively a change of duty. As a negotiator, the surveyor's duty is to his client. As an expert witness his overriding duty is to the court (CPR Part 35). The RICS is firmly of the opinion that this duty and the expert's ability to be impartial and independent are incompatible with contingency fees. Best practice guidance recommends that a surveyor notify his client that contingency fee arrangements may jeopardise the credibility of his evidence. The Practice Statement also requires the surveyor to notify the court of any issue that could affect the validity of his opinion. Contingency fees are regarded as such an issue. Whether a member of the RICS or not, valuers should heed the guidance given in Surveyors Acting as Expert Witnesses.

29.6 Immunity

At the time of writing there exists a degree of protection extended to expert witnesses not enjoyed by professionals acting directly for a client. This is referred to as immunity and in practice prevents experts from being sued for negligence as a result of their actions or inactions in undertaking an expert witness role. The original reason for immunity was that an expert may be reluctant to give evidence without such protection. The case of *Stanton* v *Callaghan* [2000] 1 QB 75 stated that the primary consideration was 'the need to ensure that the administration of justice is not impeded'.

However, valuers acting in an expert witness capacity must take care not to overlook the fact that, if considered to have been negligent, they are still liable to disciplinary procedures by their professional bodies. While there is no part of CPR Part 35 that provides for specific sanctions against an expert if he is in breach of his duty under the Rules, a judge who believes an expert has breached his duty under Part 35 is encouraged, where he feels it appropriate, to refer the matter to the expert's professional body. This was established in the case of *Pearce* v *Ove Arup Partnership Ltd* and famously tested in the case of Professor Sir Roy Meadow in reference to his evidence given in the case against Sally Clark. Meadow unwittingly provided evidence beyond his area of expertise, which misled the court and in turn resulted in the wrongful incarceration of Mrs Clark for more than two years. (The expert opinion he presented to the court relied on a statement of statistical inaccuracy — he was a paediatric consultant, not a statistician and, therefore, not qualified to make such statements). Meadow was subsequently found guilty of serious professional misconduct by his presiding professional body, the General Medical Council, and struck off the medical register. Although this is an example of flawed and ultimately misleading evidence given in a criminal case, the principle still applies in Civil Courts. The *Meadow* case highlights a stark warning to all surveyors and valuers holding themselves up as experts.

Furthermore, failure to adhere to the protocol in the case of *Philips v Symes* resulted in the award of costs against the expert personally. So, not only can providing erroneous evidence result in you being removed from your professional body, it can also cost you a hefty legal bill. Understanding not only your subject area, but also the relevant guidance is vital if an expert is not to be caught out. After all, ignorance of the law is no defence!

The subject of immunity for expert witnesses is controversial and there has been much pressure for it to be removed. Valuers by now will realise that, prior to accepting instructions to undertake valuations in court proceedings, they will need to be fully aware of the latest developments in relation to expert witness work and that any such instruction should not be taken on lightly.

An example report

(This example is given assuming a divorce, where the value of the farm occupied by both parties, until the breakdown of the marriage, is in dispute).

SINGLE JOINT EXPERT WITNESS REPORT

To:	Hereford County Court
Case No:	HR07B12345
In the Matter of:	Matrimonial Proceedings between Mary Jane Morgan and John Robert Morgan
Subject of this Report:	The Value of Broad Farm, Dunley, Worcestershire
Upon Joint Instructions from:	Graylings Solicitors 190 High Street Worcester WR1 2YJ Blacks Solicitors 164 High Street Worcester WR1 2TR
Report by:	Mr TD Jones FRICS FAAV Jones & Jones, Chartered Surveyors 281 Broad Street Hereford HR1 2TX
Date of Report:	7 July 2007

Contents

Appendices

1.0   Introduction

1.1   Instructions

Joint instructions were received from Graylings Solicitors and Blacks Solicitors in a letter dated 11th June 2007 to make an inspection of Broad Farm, Dunley, Worcestershire and provide the Court with an indication of market value of the property, assuming willing vendor and purchaser.

1.2   The Civil Procedure Rules & The Civil Justice Protocol

This report has been prepared in accordance with Part 35 of the Civil Procedure Rules (CPR), The Civil Justice Protocol for the Instruction of Experts to give Evidence in the Civil Courts, and the Royal Institution of Chartered Surveyors (RICS) Practice Statement and Guidance Note Surveyors Acting as Expert Witnesses (2nd edition), a copy of which is at Appendix 5.

1.3   The Expert Witness

My name is Thomas David Jones. I am the senior agricultural valuer and joint founding partner of the firm Jones & Jones, Chartered Surveyors in Hereford. I hold a degree in Rural Land Management from the Royal

Agricultural College, Cirencester, I am a Fellow of both the RICS and the Central Association of Agricultural Valuers. I have worked as a rural surveyor and agricultural valuer since leaving Cirencester in 1981.

2.0 Investigations

2.1 In the performance of carrying out as full an investigation as possible into the facts relevant to providing an indication of value, I have carried out the following steps:

2.1.1 An inspection of the Farm on 30th June 2007, at which Mr and Mrs Morgan were both in attendance.

2.1.2 Research into the values of comparable properties.

2.1.3 Considered market trends and factors influencing market prices.

2.1.4 Investigated the status of Town & Country Planning as it applies to Broad Farm.

2.1.5 Identified the Agricultural Land Classification, soil type, and statutory land designations.

3.0 Description

3.1 Situation

3.1.1 The property is approximately 3 miles to the southwest of Stourport-on-Severn in Worcestershire. It comprises a farmhouse, farm buildings and two main parcels of land, amounting to approximately 22.53 hectares (55 acres), on either side of the council maintained road from Dunley to Heightington (the Dunley Road). The farmstead straddles the Dunley Road on high ground in a relatively exposed position.

3.2 The Farmhouse

3.2.1 The farmhouse is constructed of brick under a concrete tile roof. Accommodation arranged over three floors briefly comprises:

- Ground floor:
  Rear kitchen, 2 reception rooms, kitchen, entrance hall, central passageway, stairs down to small cellar and stairs to first floor.

- First floor:
  Landing with doors to bathroom, 3 double bedrooms, stairs to second floor.

- Second floor:
  Landing with doors to 3 double bedrooms.

3.2.2 The house is in a poor state of internal repair and decoration. Two of the second floor bedrooms are not boarded or finished in any way, having ceiling beams and joists exposed. It is understood that the house was re-roofed in 2001. It is set in predominantly lawned gardens to the front and western side aspects with a small number of fruit trees and bordered by a hedge of native species. To the eastern and rear aspects of the house are two barns converted into residential units, understood to have been sold off from Broad Farm in the 1990s. The three dwellings are arranged in a fairly tight cluster.

3.3 **The Farmstead and Buildings**

3.3.1 The buildings are arranged on either side of the Dunley Road, with the more established farmstead being on the western side and separated from the farmhouse by one of the barn conversions.

3.3.2 The buildings on the western side of the Dunley Road comprise a Dutch Barn with lean-to to the side and rear. The main barn roof is collapsed in places. The barn is generally in a poor state of repair. The barn is set in a paddock, used as a farm yard and currently subdivided into smaller areas, understood to be used for the segregation of livestock in the normal course of farming practice. There are one or two further small barns of timber construction, largely dilapidated and not offering the possibility of any significant use. To the rear of the area housing the Dutch Barn is a separate smaller fenced area accommodating a caravan. It is understood that the ownership of this small area of land has never been disputed and forms part of Broad Farm. However, the caravan belongs to a third party who uses it for short periods from time-to-time. There is no separate access, the occupant being obliged to cross the farm yard to access the caravan plot. No further details about this were available and I was unable to obtain clearer information from either Mr or Mrs Morgan. In providing my opinion as to the value of the property, I have assumed that vacant possession would be obtainable in the event that the property was offered for sale on the open market. If this is subsequently found not to be the case, I reserve the right to amend the value stated in this report.

3.3.3 The buildings on the eastern side of the Dunley Road comprise a small collection of timber built sheds and garage, situated immediately adjacent to the gate providing access on to the Dunley Road. These buildings are not characteristically agricultural, being more of a general purpose nature and are understood to be used for the storage of miscellaneous small items.

3.4 **The Land**

3.4.1 The land is arranged on either side of the Dunley Road, and is shown in a plan at Appendix 1 together with a schedule of the field numbers and sizes. The land is arranged in two main parcels of land, extending to approximately 22.53 hectares (55 acres). The land is entirely grassland, some of which is planted to an ancient and non-productive cherry tree orchard. The ground slopes steeply to the western and eastern boundaries, making it impossible in parts to access by tractor, thereby impacting on the yield of hay and silage available. The boundaries are part fenced to a typical stock-proof standard and part enclosed by native hedgerows with some additional trees grown as wind breaks. The land is classified as grade III according to the DEFRA Agricultural Land Classification. The soil is a slightly acid loam and clay with impeded drainage. The land lies across the edge of a groundwater Nitrate Vulnerable Zone, a statutory land designation, obligating occupying farmers to adhere to certain restrictive practices in the interests of environmental protection.

3.4.2 Some of the land, notably that which is immediately adjacent to the two barn conversions and other neighbouring residential properties, may realise a higher value than that normally associated with this type of agricultural land due to its potential use as pony paddocks.

3.4.3 Broad Farm falls within the Worcestershire Unitary Authority. Under the Adopted Local Plan the property is designated as being within a Landscape Protection Area, wherein the presumption is against development likely to have a 'significant adverse effect on the quality or character' of the landscape. It seems unlikely that there would be any development possibilities at Broad Farm in the foreseeable future.

4.0 **Opinion of Value**

4.1 Having studied comparable evidence, considered market trends and relied on experience and judgement, I would estimate the value of Broad Farm to be £600,000 (Six Hundred Thousand Pounds) as at 30th June 2007.

4.2 The principle reason for the opinion reached at 4.1 is the small variation in comparable values summarised by the range of comparable properties at Appendix 6. In the experience of the Expert, comparable number 3 has the most bearing on the subject property, due to its location, character and layout. A breakdown of the various elements of value is given at Appendix 7a. These elemental values have been applied to the subject property and detailed at Appendix 7b.

5.0 **Declaration**

5.1 I understand that my overriding duty is to the Court, both in preparing reports and in giving oral evidence, and in preparing this report I have complied with this duty.

5.2 I have set out in my report what I understand to be the questions in respect of which my opinion is required.

5.3 I have done my best, in preparing my report, to be accurate and complete. I have mentioned all matters that I regard as relevant to the opinions I have expressed. All the matters on which I have expressed an opinion lie within my field of expertise.

5.4 I have drawn to the attention of the Court all material matters of which I am aware that may affect my opinion.

5.5 Wherever I have no personal knowledge, I have indicated the source of material factual information.

5.6 I have not included anything in the report that has been suggested to me by anyone, including the solicitors instructing me, without forming my own independent view of the matter.

5.7 At the time of signing the report I consider it to be complete and accurate. I will notify those instructing me if, for any reason, I subsequently consider that the report requires material correction or qualification. I understand that this report will be the evidence I will give, subject to any correction or qualification I make.

5.8 I confirm that insofar as the facts stated in my report are within my own knowledge I have made clear which they are and I believe them to be true, and that the opinions I have expressed represent my true and complete opinion

Signed . . . .
TD Jones FRICS FAAV

Dated   7th July 2007

Further Reading

Baker, Ellis, Lavers, Anthony *Case in Point — Expert Witness* (2005)

Farr, Martin *The Surveyors Expert Witness Handbook: Valuation* (2005)

The Civil Justice Council *The Civil Justice Council Protocol for the Instruction of Experts to give Evidence in the Civil Courts* (2005)

The Ministry of Justice *The Civil Procedure Rules* (1999)

The RICS Practice Statement and Guidance Note *Surveyors Acting as Expert Witnesses* (3rd edition)

The RICS *Appraisal and Valuation Standards* (6th ed)

S McLaughlin BSc(Hons), MRICS, FAAV

Glossary of General Information

30.1 Storage requirements (approximate only)

Bulk grain in store @ 16% moisture:

	m³/tonne	kg/m	cu ft/tonne
Wheat	1.29–1.35	775–741	48.4–46.3
Barley	1.44–1.50	694–667	52–54
Oats	1.96–2.07	510–483	70–74
Beans	1.21–1.27	854–816	43–46
Oil seed rape	1.40–1.46	714	50
Maize	1.32–1.38	780–748	47–50
Peas	1.29–1.35	775–741	46–48

Roots (in bulk)			
Mangolds	1.78	562	64
Swedes	1.78	562	64
Turnips	1.9	510	67
Potatoes	1.6	645	55
Fodder beet	1.78	562	64

Other commodities

Hay

Small bales	8–10	100–125	282–353

Wheat straw

Small bales	11–12	83–90	388–423

Barley straw
Small bales   13   77   459

Silage (approx only)
Wilted grass   1.25   800   44
Tower silage/haylage
 (45–50% DM)   1.55–1.77   560–640   54–62
Maize silage   1.3   770   46

Concentrates
Meal   2   500   71
Cubes/Nuts   1.6   625   56

	t/m³	cwt/cu yd
FYM (in heap)	0.93–1.06	14–16

30.2 Seeding rates

Cereal seed — commercial/dressed	Per ha	Per acre
Winter wheat	185 kg	75 kg
Winter barley	185 kg	75 kg
Winter oats	185 kg	75 kg
Spring wheat	220 kg	90 kg
Spring barley	185 kg	75 kg
Spring oats	210 kg	85 kg
Potatoes (early) (once grown)	3,700 kg	1,500 kg
Potatoes (main crop) (once grown)	2,500 kg	1,000 kg
Swedes (graded seed)	1.5 kg	0.5 kg
Mangolds (graded seed)	6.5 kg	2.6 kg
Turnips (graded seed)	1.25 kg	0.5 kg
Fodder beet (pelleted seed)		(55,000 seeds)
Stubble turnips	5–7.5 kg	2–3 kg
Rape (broadcast)	7.5–12.33 kg	3–5 kg
Kale — marrow stem (graded)	1.25 kg	0.5 kg
Kale — hybrid (graded)	1.25 kg	0.5 kg
Cabbages (graded)	0.25 kg	0.5 kg
Maize		(45,000 seeds)
Grass seeds (long leys)	33–37 kg	13–15 kg
Forage peas	86 kg	35 kg
Lucerne	25 kg	10 kg
Oilseed rape	6–8.5 kg	2.5–3.5–A kg
Field beans	250 kg	100 kg
Field peas	250 kg	100 kg

30.3 Crop yields

	Per ha	Per acre
Winter wheat	7.5–9.75 tonnes	3–4 tonnes
Spring wheat	5 tonnes	2.5 tonnes
Winter barley	6.5 tonnes	3 tonnes
Spring barley	4.33 tonnes	2 tonnes
Winter oats	7.5 tonnes	3 tonnes
Spring oats	5 tonnes	2.5 tonnes
Hay (first cut)	5–6 tonnes	2–2.5 tonnes
Grass silage (first cut)	14–20 tonnes	6–8 tonnes
Maize silage	28–34 tonnes	12–14 tonnes
Potatoes (main crop — not irrigated)	40 tonnes	18 tonnes
Swedes (precision drilled)	50–75 tonnes	20–30 tonnes
Mangolds	75–100 tonnes	30–40 tonnes
Sugar beet (washed)	40 tonnes	16–17 tonnes
Cabbages	75–100 tonnes	30–40 tonnes
Fodder beet	85 tonnes	35 tonnes

30.4 Fertilisers

Fertilisers are measured in units of N (nitrogen), P_2O_5 (phosphate), and K_2O (potash (NPK); 1 unit is the percentage of one 50kg bag of fertiliser, ie a compound, with an analysis of 20:10:10 = 20 units of nitrogen, 10 units of phosphate and 10 units of potash. Thus, if a farmer is applying 5 × 50kg of 20:10:10 per ha (2 × 50kg per acre) he is applying:

> 100 units nitrogen per ha (40 units of N per acre)
> 50 units phosphate per ha (20 units of P_2O_5 per acre)
> 50 units potash per ha (20 units of K_2O per acre).

Phosphate, for the purpose of calculating unexhausted manurial values, is further divided into soluble and insoluble, and this is invariably shown on the bag.

Some intensive farmers use a great deal of nitrogen and there are cases of as much as 400 units of nitrogen per acre (1,000 units per ha) being used by dairy farmers, at average intervals of three weeks during the growing season. Similarly, some intensive and specialist cereal growers apply three or possibly four top dressings of nitrogen at three-weekly intervals in the spring — often 170 units and more. Nitrogen is rarely applied in the winter as rainwater leaches (washes) it out of the soil. It is always wise to have the soil tested for phosphate and potash levels. Basic slag, when obtainable is very good for grassland and is generally applied at the rate of 0.5 tonne per acre (1.25 tonnes per ha). It encourages clovers, which in turn improves the fertility by naturally fixing nitrogen.

Straight fertilisers are straight nitrogen (eg Nitram), phosphate (eg Superphosphate), potash (muriate or sulphate of potash).

Compound fertilisers are compounds or combinations of these, eg 20:10:10 as described above.

Typical fertiliser applications are:

Winter corn	5–6 × 50 kg low nitrogen compounds (eg 9:25:25) per ha (2–2.5 × 50 kg per acre). Spring dressing of, say, 5 × 50kg nitrogen (eg Nitram — 34.5% N) per ha (2 × 50kg per acre). Sometimes, further nitrogen will be applied 3–4 weeks later (and possibly more) up to a total of 420 units per ha (170 units per acre).
Spring corn	5–6 × 50kg compound (eg 20:10:10) per ha (2–2.5 × 50kg per acre), together with further top dressing of nitrogen, as above.
Potatoes	1.5 tonnes per ha (0.5 tonne per acre) high potash, low phosphate compound (eg 20:8:14) before planting.
Sugar beet	Salt (eg Beetrox) in the autumn/winter at 0.5 tonne ha (4 × 50kg per acre). Before planting, 0.75 tonne compound (manufacturers have their specialist fertilisers) per ha.
Grassland	5 × 50kg grassland compound (eg 20:8:14) per ha (2 × 50kg per acre) with 3–4 further applications of nitrogen (say 50 units a time) in the season, depending on the intensity of the enterprise.

30.5 Lime

This is not a fertiliser, but is an essential nutrient and is employed to counteract soil acidity. It is necessary for nitrification, nitrogen fixation, etc. The intensity of soil acidity is measured in terms of pH units. The optimum pH for most crops is 6.5. Below this, lime requirements can vary considerably according to soil type. The heavier the soil the more lime is required, many severely deficient cases requiring an application of 7.5 tonnes per ha (3 tonnes per acre) upwards. However, it is not practicable to apply more than 7.5 tonnes per ha (3 tonnes per acre) per application. The cost in 2008 is variable, but, dependent on the amount supplied (the more supplied the cheaper it will be) and haulage costs, the average in Herefordshire is from £14 per tonne to £17 per tonne.

30.6 Sprays

There are four main types of crop protection products:

(1) herbicides
(2) fungicides
(3) insecticides
(4) growth regulators.

All the above are integral parts of any crop production system. Below are some examples of crop protection products for a number of crops (2008 costs).

(i) Winter wheat/barley

Herbicides (inc wild oats)	£50.00 per ha
Fungicides	£70.00 per ha
Insecticides	£5.00 per ha
Growth regulator	£11.00 per ha
Total	£136.00 per ha

(ii) Potatoes (Ware)

Herbicides	£45.00 per ha
Fungicides	£180.00 per ha
Insecticides	£20.00 per ha
Total	£245.00 per ha

(iii) Winter oilseed rape

Herbicides	£50.00 per ha
Fungicides	£50.00 per ha
Insecticides	£10.00 per ha
Total	£110.00 per ha

(if) Combining peas

Herbicides	£45.00 per ha
Fungicides	£30.00 per ha
Insecticides	£10.00 per ha
Total	£85.00 per ha

(v) Sugar beet

Herbicide (multi low-dose programme)	£40.00 per ha
Fungicide	£10.00 per ha
Insecticide	£15.00 per ha
Total	£65.00 per ha

30.7 Example of 2008 costs of making good dilapidations

These are intended as a guide only and are costings current in Herefordshire and Gloucestershire. VAT is also payable where applicable.

Note: These figures should be used with care having regard to all the circumstances, eg some ditches are easier to clean out than others and some hedges easier to lay.

30.7.1 Hedging and fencing

(a) Cutting and laying very high hedges including pleaching and clearing up afterwards £7–£10 per m.

(b) Cutting and laying medium and small hedges supplying stakes clearing up afterwards £4.50–£5.50 per m.

(c) Providing and planting two rows of quickthorns and protecting on both sides with pig netting and single stand barbed wire. £10–£15 per m.

(d) Trimming hedges with flail type trimmer. The cost is from £15 to £18 per hour.

(e) Siding up high hedges with flail trimmer — say 500m per hour — 3p–4p per m.

30.7.2 Fencing and gates

All erected to the workmanlike standards and all timber tanalised treated.

(a) Pig netting (medium) on tanalised posts at 2.4m centres with two strands of barbed wire £4 per m (at least 1.2m high).
(b) Post and 4 rail fencing at least 1.2m high. Sawn posts and rails tanalised £12 per m.
(c) Post and 4 rail fencing but posts round and rails half round £10.50 per m.
(d) Barbed wire fence, 2 strands, with treated softwood posts at 3m centres £3 per m.
(e) Barbed wire fence, single strand, 5m centres on treated softwood posts £2.25 per m.
(f) 3.6m galvanised medium gauge metal gate complete with hanging and shutting posts and all fittings £200 each.
(g) 3.6m oak gate with posts and all fittings £400 each.
(h) 4.5m medium gauge galvanised metal gate and posts and all fittings £250 each.

30.7.3 Ditching and drainage

(a) Ditches
Cleaning out a small ditch (by hand) up to 0.5m depth, £2.00 per m. Mechanically cleaning out silted and filled up deep ditches and depositing spoil at sides.
1.5m wide at top 2/3m at bottom — £2–£3 per m.
2.5–3m wide at top — £3.50–£4.50 per m.

(b) Land drainage
Excavate trenches, supply and lay and backfill (minimum 5ha)
160mm plastic pipe £4.20 per m run
150mm plastic pipe £4.00 per m run
100mm plastic pipe £3.25 per m run
80mm plastic pipe £3.00 per m run
60mm plastic pipe £2.00 per m run
Junction pipes £5.00 each
Backfill for 150mm pipe £1.25 per m run
60–100mm pipe £1.00 per m run
360mm pipe £1.00 per m run
Outfalls £25 each
Delivery of plant £100

(c) Dragline
Hymac £25–£30 per hour
Drott £30 per hour
Smaller machine (JCB) £20 per hour

(d) Mole drainage
£70–£75 per ha.

30.7.4 Material costs (all plus VAT)

Sheep netting £45 per roll (50m)
Pig netting (medium gauge) £28 per roll (50m)
Barbed wire (two wires) £20 per roll (200m)
Plain galvanised wire — high tensile £26 per roll (410m)
Straining posts 2.1m long tanalised
    120–150mm dia £7–£8 each
Intermediate posts 1.7m long tanalised
    75mm dia — £1.50 each
    100mm dia — £2 each
Chestnut fencing stakes £1.20 each.

Rails
3.7m long tanalised
Half round (36m × 150 mm) £5 each
Rectangular (75mm × 36mm) £3.75 each.

Gates
3.6m medium gauge primed £60 per unit
3.6m medium gauge galvanised £105 per unit
4.5m heavy gauge primed £75–£80 per unit
4.5m medium gauge primed £85 per unit
4.5m medium gauge galvanised £110 per unit
Labour for erecting gates (concreting not included) £100 per unit
Ironwork for gates £60 per unit

Gate posts
Iron gate posts (primed) £60 per unit
Ex railway sleeper posts £15–£16 per unit

30.7.5 Basic costs (2008)

Worker day £86.75 per day
Two wheel drive tractor (say 80 hp and not including implements)
£78.40 per day

30.8 Time-limits (under 1986 Act unless otherwise stated)

(1) *Claim for 1986 compensation provisions to apply to tenant right claim where tenancy commenced before 1 March 1948 — schedule 12, paras 6–9*

Tenant must elect, in writing, before tenancy terminates that compensation provisions specified in Part II of Schedule 8 of the 1986 Act shall apply to him.

(2) *Termination of tenancy claims —* landlord's and tenant's claims — section 83(2). Notice of intention to claim must be served before two months have expired from the termination date.

Eight months are allowed from the termination date to settle the claims. If not settled within this period, the matter is then referred to arbitration (sections 83(4) and (5)).

(3) *Variation of rent — section 12*

Unless agreed otherwise in the tenancy agreement, or lease, rental cannot be varied earlier than the expiry of three years from the following dates:

(i) date tenancy commenced
(ii) date of previous alteration or date an arbitrator directed that the rental should not be changed.

A notice requiring arbitration on the rental must be served in the case of a yearly tenancy, at least one year before the anniversary date of the tenancy.

(4) *Tenant's fixtures — section 10*

These can be removed at any time during the continuance of the tenancy or before the expiration of two months from the termination of the tenancy provided the tenant has paid all rent and satisfied his other obligations to the landlord and at least one month before the exercise of the right and termination of the tenancy giving the landlord written notice of his intentions to remove the fixture. The landlord can, before the expiration of the notice, give a written counter–notice electing to purchase the fixture or building at the fair value thereof to an incoming tenant.

(5) *Disturbance — section 60*
If claim is to be made for more than one year's rent, tenant must give landlord an opportunity of valuing the stock and must give written notice of intention to make such a claim not less than one month before termination of tenancy followed by a claim under section 83 within two months of the date of termination.

(6) *Deterioration — section 72*
Notice of intention to claim for deterioration (as distinct from dilapidations) must be served not later than one month before the tenancy terminates.

(7) *Claim for Tenancy Succession on Death of Tenant — Agriculture (Miscellaneous Provisions) Act 1976 Part II*
Application for tenancy must be lodged with the Agricultural Land Tribunal within three months beginning with the day after the date of death of the tenant.

(8) *Notice to Remedy Breaches — Agricultural Holdings (Arbitration Notices) Order 1987 (SI 1987 No 710)*
If a tenant wishes to contest liability to do any work specified to be done in a notice to remedy, he must serve a notice in writing on his landlord requiring arbitration within one month after the service on him of the notice to remedy.

(9) *Determination of standard milk quota and tenant's fraction before end of tenancy — Agriculture Act 1986 Part III section 10*
Landlord or tenant may at any time before the termination of the tenancy serve a written notice on the other, demanding that the determination of the standard quota for the land and the tenant's fraction shall be referred to arbitration (under the provisions of section 84 of the Agricultural Holdings Act 1986 and not under the Arbitration Act 1996).

(10) *Arbitration*

All arbitrations under the Agricultural Holdings Act 1986 (except milk quotas) are now under the provisions of the Arbitration Act 1996.

Agricultural Tenancies Act 1995

(11) *Notice to Landlord and Tenant of Farm Business Tenancy*

Both parties should give written notice to each other on or before the commencement of the tenancy identifying the land, a statement that the tenancy is to be and remain a farm business tenancy and that the character of the tenancy is primarily or wholly agricultural (section 1(4)).

(12) *Notice to terminate farm business Tenancy of more than two years*

A farm business tenancy for a term more than two years shall, instead of terminating on the term date, continue (as from that date) as a tenancy from year to year but otherwise on the terms of the original tenancy so far as applicable, unless at least twelve months before the term date a written notice has been given by either party to the other of his intention to terminate the tenancy.

(13) *Tenant's notice of intention to claim compensation for improvements*

Notice must be given within two months beginning with the date of the termination of the tenancy (section 22).

(14) *Rent variation (unless agreed otherwise in the agreement)*

Notice in writing that rental shall be referred to arbitration:

(a) at least 12 months but less than 24 months before the date of review
(b) at a date specified (or specified intervals) within the agreement
(c) if no specified date or interval, on the anniversary of the beginning of the end of the period of three years within the later of

(i) the beginning of the tenancy
(ii) the date as from which a previous determination or agreement took effect.

30.9 Metric conversion factors

METRIC TO IMPERIAL

IMPERIAL TO METRIC

Weight
1 kilogram = 2.205 lbs
1,000 kilograms = 0.984 ton
1 tonne = 1,000 kg (2,204 lbs)

Weight
1 lb = 0.454 kilogram
1 cwt = 50.80 kilograms
1 ton (2,240 lbs) = 1,016 kilograms

Length
1 millimetre = 0.04 inch
1 centimetre = 0.39 inch
1 metre = 3.279 feet
1 metre = 1.094 yards
1 kilometre = 0.6214 mile

Length
1 inch = 2.54 millimetres
1 inch = 2.54 centimetres
1 foot = 0.305 metre
1 yard = 0.914 metre

Volume
1 litre = 0.22 gallon
1 litre = 1.76 pints
1 cubic metre = 1.307 cubic yards
1 cubic metre = 35.315 cubic feet

Volume
1 gallon = 4.546 litres
1 pint = 0.568 litre
1 cubic yard = 0.765 cubic metre
1 cubic foot = 0.028 cubic metre

Area
1 hectare (10,000 m^2) = 2.471 acres
1 sq metre = 1.196 sq yards
1 sq metre = 10.764 sq feet

Area
1 acre (4,840 sq yds) = 0.405 hectares
1 sq yard = 0.836 sq metre
1 sq foot = 0.093 sq metre

Rates per acre
Conversion factor
kg/ha to lbs/acre × 0.89217
lbs/acre to kg/ha × 1.12085
kg/ha to cwt/acre × 0.0079658
cwt/acre to kg/ha × 125.5351
tons/acre to tonnes/ha × 2.510712

Advice to Young Practitioners

31.1 Auctioneering

Good auctioneers are born, but many, not so good, can improve with practice. A key to being a good auctioneer is to know the value of what you are selling.

The best auctioneers usually sell at a fast pace. Selling quickly keeps up the momentum and will keep up the trade even if it is a slow dragging one. Slow auctioneers make a bad trade even worse.

If you have been droving livestock, do not get on the plank or in the rostrum with dirty clothes, shoes and wellingtons. Change into presentable and clean attire and look smart.

An auctioneer's diction must be clear and he should not gabble incoherently.

Always start your sales on time.

If there is a dispute, do not argue, put the lot up again.

If there are queries or errors, clear these up on the day of the sale. A dissatisfied customer loses business. Remember everyone has friends and relatives and people talk, particularly country people.

Do not say 'I am selling' when you are not, as you will soon lose credibility.

Account to your clients for the proceeds of sale as soon as possible. A substantial cheque on account of a farm dispersal sale should be sent inside a week of the sale.

If auctioning a property, prepare your preamble in advance, deal with any alterations in the particulars, give possible buyers ample opportunity to look at the conditions of sale and contract, and to ask questions.

Property auctions are generally subject to a reserve. Unreasonably high reserves will often result in an unsold property. Endeavour to obtain a reasonable and fair reserve. Often a fluid reserve with discretion being given to the auctioneer is beneficial. When potential buyers know that you are selling, they will bid (as they otherwise risk losing the property) and the ultimate price will be better. Avoid saying 'the property is on the market' until you are about to knock it down.

If you are instructed to go to an auction to bid for a property — bid! If you do not bid, in the hope of negotiating a lower price when it is withdrawn, you may be disappointed. The owner may decide to keep the property or sell to someone else. A client will not thank a representative for not bidding when instructed, and losing the property. Obtain written and signed instructions to bid. If you are successful and purchase a property, do not sign the contract in your own or firm's name. If you have to sign, always name the buyer (for whom you should have had previous signed and written instructions) and indicate that you are acting as agent.

31.2 Valuations and claims

Do not settle any valuation or claim without obtaining your client's agreement. If one does so, without agreement, a negligence claim might result if the client is not satisfied with the settlement.

Frequently, in arbitration proceedings (except Presidential appointments) an arbitrator will ask a valuer (who is also acting as an advocate) for his appointment to act. This should be produced.

All valuations and claims submitted should be very carefully and thoughtfully prepared. Do not be rash in preparing claims. Where they cannot be finalised (as in the most of compulsory acquisition claims) submit an interim claim. A lesson the author had from his first employer was the comment 'how stupid you will look in a witness box being cross-examined by a clever barrister or lawyer'. Every valuation and claim made should be carefully worked out (with supporting evidence and comparables) and can be properly substantiated and negotiated.

Where claims or notices served are subject to a time-limit, do not leave such service to the last moment. Recorded delivery post is not always satisfactory. If the notice cannot be delivered the postman will leave a postcard stating that a letter/package could not be delivered and is available for collection at such and such a post office. If the proposed recipient does not collect, the claim/notice may be out of time. It is always best to pay more by using Guaranteed Special Delivery post where there is guaranteed delivery the following day, and also a receipt for delivery signed. Alternatively, have the claim/notice delivered by hand and obtain a signed receipt.

If acting as a valuer to a respondent do not try to settle at a pittance. There are many cases on compulsory acquisition etc where derisory offers have been made. Remember it is an imposition to find some body or other compulsorily acquiring part of one's property or making entry. How would you like to be at the receiving end? Always be fair and reasonable. There are, of course, cases where claimants may go over the top and submit frivolous claims or make claims that cannot be substantiated and in those cases a firm stand must be made. It is as well, in all cases, to remember the motto of the CAAV — 'do what is right come what may'. If this motto is adhered to and a reasonable attitude taken in all respects, one cannot go far wrong.

31.3 Giving evidence in arbitrations and enquiries

Do not get worried about giving evidence.

Always be calm, modest and do not appear arrogant or clever, as this may rebound on you.

Be firm and do not get flustered.

Have any notes, plans, notebooks, etc which may be needed, ready to hand.

Rely on facts and opinions that can be substantiated and do not go over the top and make wild assertions.

Do not argue with the judge or arbitrator.

When cross-examined endeavour to answer in a calm, fair and reasonable manner.

In recent years, the practice has grown (to save time, and expense) at arbitrations (and always in courts) to exchange witness statements, or proofs of evidence, including budgets, in advance of the hearing. This is to be commended as it does save considerable time at hearings and expense.

31.4 General

Always act in a straight and honest way and deal with people as you would wish to be dealt with.

Be efficient in your dealings. Deal with work promptly, correspondence on the day it is received. Do not leave until tomorrow what you can do today.

Act in a fair and reasonable matter in all your dealings. Some clients (not all fortunately) are avaricious. Try to instil some reason into those.

If a land agent, do not act in an overbearing manner towards your tenants.

Be fair to your staff. Good, reliable, efficient staff are difficult to find. Look after them or they may leave you. The replacement could be disappointing.

Start work early in the morning. Much more can be achieved in the morning than later in the day.

Be punctual.

On a personal note — always dress properly and look smart. Remember you are a professional person.

31.5 Valuation book

A tenant right or other agricultural valuation is very often prepared as one progresses through the work and entered into a valuation book. Many firms, however, have now discontinued using valuation books and prepare their valuations on loose leaf pads or folders which are then easily filed in the proper file dealing with the case and can be speedily and easily accessed.

The writer has for many years adopted this procedure rather than use valuation books and in so doing, normally serves on the opposing party a written claim, in accordance with the statute concerned. However, a candidate for the practical examination of the CAAV is required to use the valuation book that is issued at the examination. CAAV have now published suitable books for use in valuations.

The following are draft notes on the use of valuation books, issued by the CAAV.

3.5.1 Introduction

The Valuation Book is an essential tool for an agricultural valuer.

It should be a bound book — not loose leaf — with each page numbered and ruled with double cash ruling. Each book may be used for one valuation only or for several in sequence as chapters.

3.5.2 Purpose

It is used to record the items being valued in order with all appropriate notes and valuations. As a bound book, the pages are in order so that it is self-evidently a complete record which may be used in evidence at arbitration or in legal disputes. It may be relied on as evidence of matters of fact or liability many years later.

It is not an aide-memoire or a notebook and may be consulted and relied on after you have left the firm, retired or died.

31.5.3 Keeping the book

It is fundamental that the valuation book is laid out clearly, neatly and in a consistent way. Different areas and practices have varied styles in laying the book out.

A valuer with an organised book that is clearly set out will be better prepared and protected than someone shuffling loose papers. It will be a clear record in the event of any professional indemnity claims.

It should be understandable to all. The use of code or secret writing only readable by the author or his office will diminish the document's use as evidence.

All work should be in pencil.

The farm, the names of the parties and the date of the valuation are recorded at the beginning of the book.

The pages should then be used in an orderly sequence and pages numbered. Many valuers use the left page for workings and the right-hand page for the assessment of each item. In a tenant right valuation some valuers will put the items that are to the credit of the outgoing tenant on the left-hand page and those to the credit of the landlord on the right-hand page.

31.5.4 *Approach to end of tenancy valuations*

It is prudent to note:

- the repairing and other relevant terms of the tenancy agreement and
- the items to be taken over with reference numbers on the fixture notice.

Have a plan of the farm to accompany the valuation book, reconciling the field numbers if necessary.

At the first field, both valuers should record in their books:

- the parcel number(s) and area with the use and standing crop (if any)
- the cropping history together with fertiliser and lime applications. These will assist the valuation of the pasture and provide a check on the volume of lime and artificial fertiliser to be expected in the UMV claim;
- the hedges, fences and boundaries clockwise from the point of entry (this means the first fence listed may well be against the field last inspected), ascertaining which ones belong and making full notes of defects and proposed remedies, assessing their cost in money, or time (by man or machine hour or day).

Record stored and conserved crops as appropriate with weights and quantities. Make diagrams of relevant items such as silage pits, including cross-sections with measurements.

Buildings and dilapidations to them should be recorded in logical order with reference to a ground plan on which each building has a number or letter.

Record notes against each item. Remember this is the documentation explaining the detail of the handover. Record anything that will be needed in future negotiations. Bring forward to the valuation book any notes from inspections made before the harvest or ploughing but retain the original records — these inspections should have been notified to the other party.

Where the value of any item is not calculated in the field but left to be worked out later put XX in the cash column to ensure it is not missed.

Where there is a disagreement over any matter, record that fact and any offer made.

As agreement is reached, field by field, some valuers will transfer the subtotals to analysis paper so that all the information needed to prepare the valuation agreement can be consolidated, save the notes which will need to be taken from the book. The book remains as the detailed record of all the information.

31.5.5 Approach to stocktaking valuations

Record the name of the farm and the client, the balance sheet date and the date of your inspection.

In setting out the valuation book, many valuers will put all notes on the left-hand page and actual valuation totals on the right-hand pages.

Record in logical order:

- livestock by category
- severed and conserved crops by category
- goods in store
- field by field, cultivations and applied inputs.

Finish with a summary schedule to form the basis of your valuation certificate.

Index

V

W

Printed in Great Britain
by Amazon